AI in Material Science: Revolutionizing Construction in the Age of Industry 4.0

Editors

Dr. Syed Saad
Department of Civil Engineering
CECOS University of IT and Emerging Sciences
Pakistan

Syed Ammad
Department of Engineering and Built Environment
Griffith University, Gold Coast, Australia

Kumeel Rasheed
Department of Computer Engineering
Bahria University, Pakistan

CRC Press is an imprint of the
Taylor & Francis Group, an **informa** business

First edition published 2025
by CRC Press
2385 NW Executive Center Drive, Suite 320, Boca Raton FL 33431

and by CRC Press
4 Park Square, Milton Park, Abingdon, Oxon, OX14 4RN

CRC Press is an imprint of Taylor & Francis Group, LLC

Library of Congress Cataloging-in-Publication Data (applied for)

ISBN: 978-1-032-56932-1 (hbk)
ISBN: 978-1-032-57225-3 (pbk)
ISBN: 978-1-003-43848-9 (ebk)

DOI: 10.1201/9781003438489

Typeset in Times New Roman
by Prime Publishing Services

Preface

As an academic exploration, this book navigates the ever-evolving landscape of Artificial Intelligence's (AI) influence on modern materials within the Industry 4.0 era. The comprehensive nature of this work is rooted in the foundational principles and realistic applications of AI, focusing on its integration into material science and construction practices. Our collective goal in crafting this collection stems from a recognition of the widening breach between academic literature and industry-centric publications within the AI and materials science domain. AI has become an industry-wide revolutionary force in the context of Industry 4.0. Its impact goes beyond the Internet of Things' integration to include the adoption of creative healthcare solutions. Specifically, materials science is one area where artificial intelligence is very important for advancing material innovation as this is a broad topic that includes design, performance prediction, and synthesis. AI is projected to significantly transform the process by which new materials are planned, produced, and developed as it continues to advance in the field of materials science.

Primarily, this book attempts to bridge the gap by offering a refinement perspective that caters to both scholarly intellects and industry professionals. The intention is twofold: Firstly, to provide a modern reference that extends the existing discourse, surpassing the temporal constraints of earlier seminal works. Secondly, it aspires to summarize the advancements in AI, and search through into complex features from AI-driven material optimization to its role in the augmentation of construction technologies. This book focuses the AI's influence on metallic work materials and chip-forming cutting processes, finding a way around extensive subjects like machine tool control. Within materials science, AI presents scientific, technical, and social opportunities, emphasizing sustainable material development. Researchers' and engineers' capacities have been greatly improved by the combination of AI and materials science. Artificial Intelligence (AI) facilitates the design of new materials with customised features by using advanced computations and machine learning simulations, revealing up previously unexplored possibilities. Moreover, AI's predictive powers advance our knowledge of material behaviour and make it possible to project performance in a variety of scenarios with greater accuracy.

We express our sincere appreciation to all the authors who contributed their expertise and insights to this book. Their devotion and hard work have enhanced the content, transforming it into a thorough and valuable resource. Each author's unique perspective has played a vital role in the success of this collaborative endeavor. Lastly, we are incredibly grateful to our parents and spouse, i.e. Syed Manzoor Ali Shah, Azra Yasmine, Ume Hani Naeem, and Muhammad Sadiq, Rohina Sadiq, who have been the solid foundation of our journeys with their steadfast support and endless encouragement. Their love and guidance will always be remembered.

Dr. Syed Saad
Syed Ammad
Kumeel Rasheed

Contents

Abbreviations

AR : Augmented Reality
VR : Virtual Reality
CPS : Cyber-Physical System
BIM : Building Information Modelling
IoT : Internet of Thing
CNTs : Carbon Nanotubes
AI : Artificial Intelligence
ANN : Artificial Neural Network
HVAC : Heating, Ventilation, and Air Conditioning
MIT : Massachusetts Institute of Technology
RPA : Robotic Process Automation
BD : Big Data
NLP : Natural Language Processing
CNNs : Convolutional Neural Network
XAI : Explainable Artificial Intelligence
GPUs : Graphics Processing Units
TPUs : Tensor Processing Units
SCC : Self-Compacting Concrete
LiDAR : Light Detection and Ranging
LNG : Liquefied Natural Gas
HPC : High-Performance Concrete
SVM : Support Vector Machines
BRT : Bagging Regression Trees
MART : Multiple Additive Regression Trees
SPD : Surge Protective Device
SCM : Supplementary Cementing Materials
FEM : Finite Element Method
BPNN : Back-Propagation Neural Networks
RF : Random Forest
RL : Reinforcement Learning
RtL : Real-time Learning
BAS : *Beetle Antennae Search*

CBR	:	California Bearing Ratio
ANFIS	:	Adaptive Neuro-Fuzzy Inference
ML	:	Machine Learning
GA	:	Genetic Algorithms
MFA	:	Modified Firefly Algorithm
TSR	:	Tensile Strength Ratio
GEP	:	Gene Expression Programming
FRPs	:	Fiber-Reinforced Polymers
OSB	:	Oriented Strand Board
SMPs	:	Shape Memory Polymers
CAD	:	Computer-Aided Design
CAM	:	Computer-Aided Manufacturing
HRI	:	Human-Robot Interface
HCGR	:	Hybrid Motioned Curtain-Wall Glazing Robot
GCGR	:	Glass Ceiling Glazing Robot
CSIC	:	Construction Scotland Innovation Centre
RCSS	:	Robotized Construction Site Surveyor
RVIO	:	Robust Visual-Inertial Odometry
GB	:	Graph Based
SIP	:	Structural Inspection Planner
AGP	:	Art Gallery Problem
TSP	:	Travelling Salesman Problem
UAV	:	Unmanned Aerial Vehicles
UGV	:	Unmanned Group Vehicles
GPU	:	Graphic Processing Unit
DFMA	:	Design for Manufacture and Assembly
IDP	:	Intelligent Document Processing
OSHA	:	Occupational Safety and Health Administration
DL	:	Deep Learning
RL	:	Reinforcement Learning
KR&R	:	Knowledge Representation and Reasoning
AGI	:	Artificial General Intelligence
KNN	:	K-Nearest Neighbor
GPSR	:	Genetic Programming-Based Symbolic Regression
NN	:	Neural Network
OCR	:	Optical Character Recognition
DSM	:	Design Structure Matrix
AEC	:	Architecture, Engineering, and Construction
SD	:	Schematic Design
DD	:	Design Development
PD	:	Pre-Design
CD	:	Construction Documents
CA	:	Construction Administration
AM	:	Additive Manufacturing

CNC	:	Computer Numerical Control
GA	:	Genetic Algorithms
AIS	:	Artificial Immune Systems
ACO	:	Ant Colony Optimization
PSO	:	Particle Swarm Optimization
ISF	:	Incremental Sheet Forming
ICT	:	Information and Communication Technology
MiC	:	Modular Integrated Construction
GDP	:	Gross Domestic Product
USD	:	United States Dollar
ETA	:	Estimated Time of Arrival
GPS	:	Global Positioning System
CO_2	:	Carbon Dioxide
WHO	:	World Health Organization
Covid-19	:	Coronavirus Disease 2019
GANs	:	Generative Adversarial Networks
2D	:	Two-Dimensional
3D	:	Three Dimensional
RFID	:	Radio-Frequency Identification
OECD	:	Organization for Economic Cooperation and Development
MDPI	:	Multidisciplinary Digital Publishing Institute
IEEE	:	Institute of Electrical and Electronics Engineers
ITU	:	International Telecommunication Union
SDGs	:	Sustainability Development Goals
ACM	:	Association for Computing Machinery
CPCB	:	Central Pollution Control Board
UN	:	United Nation
SMEs	:	Small and Medium-Sized Enterprises
APIs	:	Application Programming Interfaces
GDPR	:	General Data Protection Regulation
HIPAA	:	Health Insurance Portability and Accountability Act

CHAPTER

1

Is AI the Architect of Tomorrow's Materials in the Age of Industry 4.0?

Syed Saad, Kumeel Rasheed, Syed Ammad, Muhammad Waqas Khan and Ahmad Zaland

Overview

The creation of new projects has been made possible by the way that materials are now developed, tested, and generated. Artificial intelligence (AI) is contributing in the speeding of the material research and development process by analyzing enormous volumes of data, simulating experiments, and predicting the properties of innovative materials. The process, which used to take years, can now be finished in weeks or even days. AI algorithms have been used to forecast the best elemental and processing conditions for developing novel alloys with higher strength and corrosion resistance. It is also being utilized in the manufacturing process to increase quality control and problem detection. AI systems can spot problems in real-time by analyzing photos and data from sensors, allowing for speedy fixes, and decreasing waste. AI-powered cameras may identify microcracks in concrete structures, preventing costly repairs and maintaining the building's safety. Predictive maintenance is another way AI is altering construction. AI algorithms can forecast when components will break by analyzing data from sensors deployed in buildings and infrastructure, allowing for preventive maintenance and reducing downtime. This technique is already in use in wind turbines, where sensors detect vibration and temperature changes that warn of a component's impending breakdown. Technicians can be warned before a failure occurs with predictive maintenance, eliminating the need for costly emergency repairs, and prolonging the turbine's lifespan. AI is assisting in the improvement of energy efficiency in buildings by analyzing sensor data and altering heating, ventilation, and air conditioning (HVAC) systems and lights accordingly. "AI algorithms can optimize energy usage by taking into account a variety of factors, such as occupancy, weather, and building orientation." This technology is already being utilized in smart buildings, where sensors detect occupancy and adjust lighting and heating as needed. These buildings may save money and minimize their carbon impact by reducing their energy consumption. This book will provide you with a complete grasp of the potential of AI in material science and how it can be utilized to revolutionize the construction industry, whether you are a researcher, engineer, architect, or construction professional.

Purpose and Scope

A technological revolution is about to occur as the world moves faster towards AI, a revolution that will permanently change how we interact, live, and work. At the forefront of this transition is the use of AI in Material Science, a field that has recently undergone a revolution thanks to extraordinary developments in the creation of AI tools and algorithms. An interdisciplinary field of study known as material science examines the composition, structure, and synthesis of materials. The potential for AI in material science is vast, and its applications are numerous. Some of the significant fields where AI is contributing significantly are the ones listed below. To design and synthesize new materials with superior properties, AI methods are being used which can assist scientists to predict the characteristics of new substances and speed up the discovery process.

1. Material Characterization: Analyzing and characterizing materials also use machine learning (ML). Data analysis from several sources, including X-ray diffraction, electron microscopy, and spectroscopy, can help scientists learn concerning the structure and characteristics of a material. AI applications in the area of materials science are already having a significant impact. Here are a few of these instances:

(a) Creating Better Batteries: A new type of battery that can function for a longer amount of time as well as recharge more quickly was created by researchers at Stanford University by utilizing ML algorithms. The research team was able to identify novel materials that may be used to boost battery capacity by analyzing data acquired from countless battery testing.

(b) Making Solar Cells Better: AI was used by MIT researchers to develop a new type of solar cell that is more effective than conventional silicon-based cells. By examining data from millions of simulations, the team was able to identify novel materials that could be utilized to improve the efficiency of solar cells. The impact of AI on the subject of material science as well as how it is changing the construction industry. Let's explore this intriguing subject a little more.

This book's goal is to offer a comprehensive analysis of how material science and AI interface in the building industry. It seeks to shed light on the most recent innovations and advancements in this field and how they are altering the construction sector.

2. Material Science: This book will cover a wide range of topics related to AI in material science, including the use of AI in construction planning and design, the use of AI-powered sensors and monitoring systems on building sites, and the role of AI in material discovery and optimization. It will address the ethical and societal ramifications of such developments, as well as the potential advantages and disadvantages of integrating AI into the construction process. Readers will come across practical instances as well as case studies that show the useful applications of AI in the construction industry and how these uses are influencing the future of this industry throughout the book. Researchers, academics, and professionals interested in the relationship between AI and materials should read this book.

In conclusion, AI is revolutionizing Material Science and has a huge impact on the building industry. In the field of material science, AI has a tremendous potential, from creating more durable substances to enhancing solar cells and batteries. The application of AI in materials science will grow as we go towards a more connected world, and in the coming years, we may anticipate more creative discoveries in the field.

Target Audience

The book has a specific audience in mind: a broad spectrum of readers interested in the nexus between AI material science, and the construction industry. Professionals, researchers, educators, students, and technology enthusiasts who are interested in learning more about how AI could revolutionize the construction industry is the book's intended audience.

Professionals in the Construction Industry

This book is intended for professionals working in the construction industry, including architects, civil engineers, managers of construction, and project managers. They gain a thorough grasp of how AI may enable smart buildings, increase structural integrity, optimize material choices, and improve construction procedures. It can help professionals stay on the cutting edge of technological developments in their field by promoting creativity and competition. Professionals in these sectors must keep up with the most recent developments and comprehend how to use AI to better material research, design, and building processes as AI-driven technologies alter the construction industry (Ammad et al., 2021). Students can improve their professional skills thanks to the book's practical advice, case studies, and best practices.

Academics and Researchers

The thorough examination of AI applications for building will be helpful to academics and researchers in the fields of material science, civil engineering, computer science, and related fields. It functions as a reference manual, providing in-depth insights into recent research, case studies, and new trends. By bridging the gap between material science and AI, the book also promotes multidisciplinary cooperation, opening up new horizons for investigation and innovation. By examining the merging of AI, material science, and building within the context of AI, this book fills an essential a vacuum in the structure of existing literature. It advances academic knowledge by addressing this specialized topic and provides researchers, academics, and learners in the field with valuable resource.

Interdisciplinary Collaboration

Participation and collaboration among experts, researchers, and specialists from several disciplines or fields of study are referred to as interdisciplinary collaboration which is essential for fostering cross-pollination of ideas, advancing innovation, and addressing difficult problems at the nexus of material science and AI. Interdisciplinary collaboration enables experts from several fields to share their own viewpoints, approaches, and information, bridging knowledge gaps (Bremer & Meisch, 2017). Cooperation between experts in the domains of materials science and AI allows for the sharing of ideas, theories, and methods that might not been previously considered. Professionals can find creative approaches and responses to the problems the construction sector faces by pooling their skills.

Cross-Fertilization of Ideas: When experts from many fields work together, there is a chance for cross-fertilization of concepts. For instance, material scientists can contribute their knowledge of the characteristics, functionality, and performance of building materials, while professionals in AI can share their knowledge of the creation of clever algorithms, ML, and data analysis. Collaboration between material scientists and AI specialists has the potential to have synergistic effects and to promote innovation. New opportunities for optimizing material qualities, forecasting performance, and creating sustainable construction solutions emerge when the computational power of AI merges with the domain-specific expertise of material science (Pilania, 2021). The exploration of unchartered waters and the encouragement of imaginative thinking are fostered through interdisciplinary collaboration, which ultimately leads to advancements across both areas.

Educational Institutions

Future professionals in the domains of material science, engineering, and construction will be shaped by educational institutions in terms of their knowledge, abilities, and competencies. For various reasons, it is essential that educational institutions work with the book's intended readership. The content can help educational institutions that offer courses and programmers in material science, engineering, and construction enhance their curricula. It also offers insights into how AI might be applied to materials science within the building sector, in line with the areas' changing needs and trends. By updating their course materials, professors and lecturers may make sure that their students are exposed to the most recent advances and cutting-edge studies. For students working on final assignments or research projects as part of material science, the field of engineering, and construction programmers, the book is an invaluable resource. It offers a thorough grasp of how AI and material science interact, giving students a solid platform on which to build creative research topics and useful project ideas. Institutions of higher learning may motivate students to use the book's lessons to further their own research and advance the body of information within the topic. To close the knowledge gap between academia and practical application, educational institutions frequently work with industrial partners. The book can act as a link, enabling fruitful partnerships between academic institutions and business parties. It may assist educational institutions better match their curricula to the demands of business, so that graduates have the expertise and skills needed to traverse the convergence of AI, material science, and construction.

Key Takeaways

In AI in Material Science, it is crucial to establish the important points that will capture readers and highlight the significance of the book's subject matter in the first chapter. The examination of AI's transformative role in the field of materials science and its tremendous impact on the building industry is laid out in this section, which also serves as the book's firm foundation.

Exploiting AI's Potential in Material Science, it generates previously unthinkable prospects for advancement and creativity through the integration of AI technology with material science research. This book explores how AI techniques, neural networks, and data analytics have the potential to revolutionize the way we analyze and design building materials, resulting in improved performance, sustainability, and efficiency.

Bringing theory and practice together, the book's main objective is to close the gap between theoretical ideas and real-world implementations. We demonstrate how AI-based solutions may be successfully applied in material science research through practical instances and examples, empowering engineers, architects, and construction professionals to make educated decisions, optimize processes, and generate cutting-edge projects.

It is of utmost importance to embrace the potential of AI in the field of material science, marked by the integration of digital technology and physical systems. To create a more sustainable and cutting-edge construction industry, this book examines how AI-driven approaches might improve production procedures, enable predictive maintenance, and accelerate up the creation of intelligent building materials.

Consequences for the future, the impacts of AI on material science extend much beyond today. We can advance the construction industry significantly by better understanding the abilities of AI and utilizing its potential. As AI keeps affecting the future of materials science in construction, this book not only analyzes current developments but also provides a forward-looking perspective, examining the moral implications, difficulties, and possibilities that lie ahead.

Fundamentals of AI

The book presents a solid case for continuing the investigation of AI in the field of material science through this academic and professional introduction, opening the door for readers to learn about the revolutionary possibilities that lie at the intersection of AI and the construction industry.

Learning about Artificial Intelligence, the fundamental concepts and tenets of AI are covered in this section. It examines the idea of AI, the distinctions between specific and general AI, and the many methods and algorithms used in systems that use AI. Readers can comprehend AI's possible uses and consequences in the setting of material science by developing a solid comprehension of the technology.

AI represents one of the most transformative and interesting fields in ultramodern technology and computer wisdom. At its core, AI is the pursuit of creating machines and computer systems that can mimic mortal intelligence, enabling them to learn, reason, problem-break, and generate opinions (Lee et al., 2019). The primary thing of AI is to replicate mortal-like cognitive capacities in machines, allowing them to perform tasks that generally bear mortal intelligence, similar as understanding natural language, feting patterns, and conforming to new information (Cope & Kalantzis, 2022). AI encompasses a wide range of subfields, including machine literacy, neural networks, natural language processing (NLP), robotics, and expert systems, all aimed at advancing the capabilities of artificial systems (Kumar & Thakur, 2012).

One of the foundational generalities in AI is machine literacy, which empowers computers to learn from data and upgrade their performance without being explicitly programmed (Liaskos et al., 2022). Machine literacy algorithms can fete patterns and make prognostications or opinions grounded on training data, making them particularly important in colorful operations like image recognition, speech recognition, and recommendation systems. Another essential aspect of AI is neural networks, inspired by the structure of the mortal brain. These networks correspond of connected bumps that reuse information in layers, enabling deep literacy models to handle complex tasks like image bracket and language restatement with remarkable delicacy.

As part of AI, interpreting and producing language becomes possible through NLP by bridging the gap between humans and machines (Rasheed et al., 2022). NLP has enabled more fluid communication through enhanced chatbot, virtual assistant, and language translation technologies. As AI penetrates different sectors and aspects of our lives, grasping its basis is crucial to exploit its potential both morally and pragmatically (Gill & Kaur, 2023). Further, the key concepts of fundamentals of AI are discussed in details.

Machine Learning (ML)

ML is a subset of AI that focuses on creating algorithms and models that can learn from data and make predictions or decisions without being explicitly programmed (Abdul Hannan Qureshi et al., 2020). There are several key components within ML.

Supervised Learning: In supervised learning, a model is trained on a labeled dataset, where the input data and their corresponding correct outputs are provided. The model learns to map inputs to outputs, making it capable of making predictions on new, unseen data.

Unsupervised Learning: Unsupervised learning involves finding patterns or structure in data without labeled outputs. Clustering, which groups related data points, and dimensionality reduction, which simplifies complicated data, are common approaches.

Reinforced Learning: The process of reinforcement learning is a sort of AI in which an agent finds out how to choose a series of actions in a setting to maximize a cumulative reward. It's widely used in robotics, game playing, and autonomous systems.

Neural Networks

Neural networks are computational models inspired by the structure of the human brain. They consist of interconnected nodes (neurons) organized in layers. Key components of neural networks include:

Input Layer: This layer receives the initial data or features.

Hidden Layers: These intermediate layers process and transform the data through weighted connections and activation functions.

Output Layer: The output layer produces the final prediction or decision.

Neural networks, particularly deep neural networks, have been instrumental in tasks like image and speech recognition, NLP, and reinforcement learning (Maind & Wankar, 2014).

Data

Data is the core component of AI. It might be unstructured (like text, photos, or videos) or organized such as databases, spreadsheets. High-quality data is critical for training accurate AI models. Data-related concepts include:

Data Collection: Gathering data from various sources, including sensors, databases, and the internet.

Data Preprocessing: Cleaning, transforming, and organizing data to make it suitable for AI model training.

Feature Engineering: Selecting or creating the most relevant features (input variables) for AI models (Roh et al., 2019).

Algorithm

A well-defined collection of instructions or rules called an algorithm must be followed in order to complete a certain task or solve a particular issue. Algorithms are crucial elements that control how AI systems analyze data, make choices, and learn from experience (Sarker, 2021).

Key Characteristics

Input: Algorithms take input data or information as their starting point. This input can be in various forms, such as numbers, text, images, or sensor readings, depending on the problem being addressed.

Output: Algorithms produce output data or a solution based on the input and the defined steps. The output can also take various forms, such as a numerical result, a decision, or a recommendation.

Effectiveness: Algorithms must be effective, meaning they solve the problem correctly according to their intended purpose. They need to offer a solution that makes sense and relates to the issue at hand.

Algorithm determinism: An algorithm will always generate the same result if its input and beginning circumstances are the same. The dependability of AI systems depends on this predictability.

Termination: Algorithms must eventually halt and produce an output. While some algorithms may take a long time to complete for complex tasks, they should not run indefinitely.

Types of Algorithms

Depending on their usage and functionality, many types of algorithms are utilized in AI and computer science.

Search Methods: These techniques allow you to find a specific item in a list or the shortest path through a network.

Sorting Algorithms: Elements are arranged according to an ordered process, like ascending or downward direction, through sorting algorithms. Bubble sort, merge sort, and quicksort are just three examples of common sorting algorithms.

Machine Learning Algorithms: ML and AI rely heavily on these algorithms. All these fall under their category: supervised, unsupervised, and reinforcement learning algorithms like linear regression, support vector machines, k-means clustering, and Q-learning.

Graph Algorithms: Algorithms for graphs work by leveraging network connections, providing effective solutions to related issues. Among examples of algorithms, you will find Dijkstra's for shortest paths and Kruskal's for building minimum spanning trees.

Genetic Algorithms: Algorithms based on genetics are a type of optimization techniques influenced by evolutionary processes. These tasks are some examples of what they are commonly used for.

Image Processing Algorithms: Algorithms used in computer vision and image analysis help process and evaluate visual data. Edge detection algorithms and image segmentation techniques are some examples of image processing algorithms.

Role of Algorithms in AI

Heart of AI systems, algorithms play a crucial role in ML and deep learning. In AI, algorithms are responsible for:

Learning: Models learn from data through algorithms in ML, enabling predictions and decisions based on patterns and relationships.

Inference: AI systems can quickly make decisions or estimates using the acquired information or patterns through the use of inference algorithms in real-time.

Optimization: Fine-tuning model parameters is precisely the task of optimization algorithms, producing AI systems that function optimally and efficiently.

Data Processing: By means of manipulating text, images, and numbers, algorithms identify important details.

Pattern Recognition: In computer vision and NLP, AI algorithms are engineered to identify intricate patterns and systems.

Developers and researchers crafting AI systems rely on algorithms, their core intelligence property, to dictate functionality, accuracy, and performance (Schork, 2019). Optimizing algorithms and choosing them is critical when developing successful AI solutions.

Natural Language Processing (NLP)

NLP is a subfield of AI that focuses on enabling computers to understand, generate, and interact with human language. Key NLP tasks include:

Speech Recognition: Converting spoken language into text.

Machine Translation: Translating text from one language to another.

Sentiment Analysis: Determining the sentiment or emotional tone of text data.

Chatbots and Virtual Assistants: Creating AI systems that can engage in natural language conversations with users (Zhou et al., 2020).

Computer Vision

With computer vision, machines may decipher and understand the visible data from the world, which includes images and videos. Human visual perception is the area of research that requires the creation of algorithms and models. Key components and concepts within computer vision include:

Image Classification: Image classification means assigning labels or categories to pictures. Accurate recognition of objects, scenes, and patterns is now possible thanks to the revolutionary use of convolutional neural networks (CNNs).

Object Detection: Classifying objects, location and identification play crucial roles in object detection. These applications often use it, and rightly so.

Image Segmentation: Dividing an image into separate regions or segments is the process of image segmentation. Medical image analysis requires precise identification of structures, and this technique is vital.

Facial Recognition: Those features allow facial recognition systems to identify and confirm individuals. With applications ranging from security to personalization, they offer.

Pose Estimation: Pose estimation helps identify the location and angle of objects or body parts in images or videos. Applications range from sports analysis to robotics and augmented reality.

Video Analysis: Object tracking, action recognition, and motion pattern analysis are possible with video data via computer vision.

3D Vision: Understanding the depth of 2D images or video requires 3D computer vision's skills. This is valuable in robotics, augmented reality, and autonomous navigation.

Deep Learning in Computer Vision: Deep learning, CNNs being a specific subset, has been central to computer vision advancements. Computer vision applications now become simpler with the aid of pre-trained deep learning models (Joby, 2021).

Expert Systems

Expert systems are AI programs designed to emulate the decision-making abilities of human experts in specific domains (Duan et al., 2019). They use knowledge representation, inference engines, and rule-based reasoning to provide expert-level advice and solutions. Here are some key components and characteristics of expert systems:

Knowledge Base: Expert systems have a knowledge base that stores facts, rules, and domain-specific information. This knowledge is typically acquired from human experts and domain literature.

Inference Engine: The inference engine is the core of an expert system. It processes the knowledge in the knowledge base and uses it to answer user queries or make decisions. It can employ various reasoning techniques, including forward chaining and backward chaining.

User Interface: Expert systems often have user-friendly interfaces that allow users to interact with the system, input data, and receive recommendations or solutions. These interfaces can be text-based or graphical.

Explanation Facility: Many expert systems provide explanations for their decisions. This transparency helps users understand the reasoning behind the system's recommendations.

Examples of Expert Systems: Expert systems have been applied in various fields, including medical diagnosis financial analysis, troubleshooting technical problems, and providing legal advice.

Limitations: While expert systems excel at replicating domain-specific expertise, they have limitations. They may not handle uncertainty or new, unforeseen situations well. Maintaining and updating the knowledge base can also be labor-intensive.

Expert systems continue to be valuable tools in fields where expert knowledge is critical, aiding decision-making and problem-solving by providing consistent, expert-level guidance (O'Leary, 2021).

Ethical Considerations

Ethical considerations are paramount in AI development and deployment to ensure responsible and fair use. Some key aspects include:

Bias and Fairness: AI systems can inherit biases from training data. It's crucial to detect and mitigate biases to prevent discriminatory outcomes in areas like hiring, lending, and law enforcement.

Transparency: AI models should be transparent, allowing users to understand how decisions are made. Explainable AI (XAI) techniques aim to provide insights into the reasoning behind AI predictions.

Accountability: Establishing accountability for AI decisions is essential. This involves identifying responsible parties and mechanisms for recourse in case of AI errors or misuse.

Privacy: AI systems often deal with sensitive data. Safeguarding user privacy is vital, with techniques like data anonymization and privacy-preserving ML.

Regulations and Standards: Many countries and regions are introducing regulations governing AI use. Adhering to these regulations is necessary for legal compliance.

Ethical Frameworks: Adopting ethical frameworks, such as the principles outlined in the Fairness, Accountability, and Transparency in Machine Learning community, can guide responsible AI development.

Hardware and Computing Resources

AI often requires specialized hardware and ample computing resources, including:

Graphics Processing Units (GPUs): GPUs are well-suited for training DL models due to their parallel processing capabilities. They accelerate tasks like image and speech recognition.

Tensor Processing Units (TPUs): TPUs are custom-designed AI accelerators developed by Google. They excel in executing ML workloads efficiently.

Cloud Computing: Cloud platforms provide scalable infrastructure for AI development, offering services for training, inference, and deploying AI models without the need for extensive on-premises hardware.

Edge Computing: In some applications, AI processing occurs at the edge (close to the data source), reducing latency and improving real-time decision-making in devices like internet of things (IoT) sensors and autonomous vehicles.

Problem Solving and Decision Making

AI is applied to solve complex problems and make decisions based on data and logic. Key techniques and concepts include:

Decision Trees: Decision trees are used for classification and decision-making. They represent decisions and their possible consequences in a tree-like structure.

Optimization Algorithms: From a pool of potential solutions, optimization algorithms choose the best one. They are employed in disciplines such as logistics, supply chain management, and resource allocation.

Reinforcement Learning: Reinforcement learning is employed for training agents to make sequences of decisions to maximize a cumulative reward. It's used in robotics, gaming, and autonomous systems.

Validation and Testing

Rigorous testing and validation processes are crucial to ensure the reliability and accuracy of AI systems. Key considerations include:

Training and Test Datasets: Models should be evaluated on separate datasets not used during training to assess their generalization ability.

Cross-Validation: Cross-validation techniques help assess model performance by dividing data into multiple subsets for training and testing, reducing the risk of overfitting.

Benchmarking: Comparing AI models against established benchmarks and industry standards helps assess their quality and performance.

Decision-making and Problem-solving: AI is used to tackle complicated issues and arrive at conclusions using logic and data. Important methods and ideas include:

Decision Trees: Classification and decision-making are accomplished using decision trees. They display choices and potential outcomes in a tree-like framework.

Optimization Algorithms: These algorithms choose the optimal option from a set of probable solutions. They work in fields including resource allocation, supply chain management, and logistics.

Reinforcement learning is used to educate agents to make decisions in a sequence that maximizes a cumulative reward. It is utilized in autonomous systems, gaming, and robotics.

References

Abdul Hannan Qureshi, Wesam Salah Alaloul, Manzoor, B., Musarat, M.A., Saad, S. and Ammad, S. (2020). Implications of Machine Learning Integrated Technologies for Construction Progress Detection under Industry 4.0 (IR4.0). *2020 Second International Sustainability and Resilience Conference: Technology and Innovation in Building Designs, 0*. https://doi.org/10.1109/IEEECONF51154.2020.9319974.

Ammad, S., Saad, S., Bashir, M.T., Qureshi, A.H., Altaf, M. and Rasheed, K. (2021). Construction Accidents via Integrating Building Information Modelling (BIM) with Emerging Digital Technologies: A Review. *2021 3rd International Sustainability and Resilience Conference: Climate Change*. https://doi.org/10.1109/IEEECONF53624.2021.9668066.

Bremer, S. and Meisch, S. (2017). Co-production in climate change research: Reviewing different perspectives. *WIREs Climate Change*, 8(6), e482. https://doi.org/https://doi.org/10.1002/wcc.482.

Cope, B. and Kalantzis, M. (2022). Artificial intelligence in the long view: From mechanical intelligence to cyber-social systems. *Discover Artificial Intelligence*, 2(1), 13. https://doi.org/10.1007/s44163-022-00029-1.

Duan, Y., Edwards, J.S. and Dwivedi, Y.K. (2019). Artificial intelligence for decision-making in the era of Big Data: Evolution, challenges, and research agenda. *International Journal of Information Management*, 48, 63–71. https://doi.org/https://doi.org/10.1016/j.ijinfomgt.2019.01.021.

Gill, S.S. and Kaur, R. (2023). ChatGPT: Vision and challenges. *Internet of Things and Cyber-Physical Systems*, 3, 262–271.

Joby, A. (2021). *Computer Vision: How Machines Interpret the Visual World*. https://learn.g2.com>computer- vision.

Kumar, K. and Thakur, G.S.M. (2012). Advanced applications of neural networks and artificial intelligence: A review. *International Journal of Information Technology and Computer Science*, 4(6), 57.

Lee, J., Suh, T., Roy, D. and Baucus, M. (2019). Emerging technology and business model innovation: The case of artificial intelligence. *Journal of Open Innovation: Technology, Market, and Complexity*, 5(3), 44.

Liaskos, C., Tsioliaridou, A., Georgopoulos, K., Morianos, I., Ioannidis, S., Salem, I., Manessis, D., Schmid, S., Tyrovolas, D., Tegos, S.A., Mekikis, P.-V., Diamantoulakis, P.D., Pitilakis, A., Kantartzis, N.V., Karagiannidis, G.K., Tasolamprou, A.C., Tsilipakos, O., Kafesaki, M., Akyildiz, I.F., ... Spais, I. (2022). XR-RF imaging enabled by software-defined metasurfaces and machine learning: Foundational vision, technologies and challenges. *IEEE Access*, 10, 119841–119862. https://doi.org/10.1109/ACCESS.2022.3219871.

Maind, S.B. and Wankar, P. (2014). Research paper on basic of artificial neural network. *International Journal on Recent and Innovation Trends in Computing and Communication*, 2(1), 96–100.

O'Leary, D.E. (2021). *Expert Systems – History, Structure, Definitions, Characteristics, Life Cycle, and Applications*. Marshall School of Business, University of Southern California.

Pilania, G. (2021). Machine learning in materials science: From explainable predictions to autonomous design. *Computational Materials Science*, 193, 110360. https://doi.org/https://doi.org/10.1016/j.commatsci.2021.110360.

Rasheed, K., Ammad, S., Said, A.Y., Balbehaith, M., Oad, V.K. and Khan, A. (2022). Application of Smart IoT Technology in Project Management Scenarios. *2022 International Conference on Data Analytics for Business and Industry, ICDABI 2022*. https://doi.org/10.1109/ICDABI56818.2022.10041674.

Roh, Y., Heo, G. and Whang, S.E. (2019). A survey on data collection for machine learning: A big data-AI integration perspective. *IEEE Transactions on Knowledge and Data Engineering*, 33(4), 1328–1347.

Sarker, I.H. (2021). Machine learning: Algorithms, real-world applications, and research directions. *SN Computer Science*, 2(3), 160.

Schork, N.J. (2019). Artificial intelligence and personalized medicine. *In:* D.D. Von Hoff and H. Han (Eds.), *Precision Medicine in Cancer Therapy*, pp. 265–283S. Springer International Publishing. https://doi.org/10.1007/978-3-030-16391-4_11.

Zhou, M., Duan, N., Liu, S. and Shum, H.-Y. (2020). Progress in neural NLP: Modeling, learning, and reasoning. *Engineering*, 6(3), 275–290.

CHAPTER

2

History of AI

Kumeel Rasheed, Ahmad Zaland, Syed Saad, Syed Ammad and Ali Rostami

Introduction

Artificial intelligence (AI) has been investigated for years, and is still among the most enigmatic areas in computer wisdom. This is incomplete because the issue is so broad and hazy. AI includes everything from extremely intelligent robots to search algorithms used in board games. It has applicability for almost all societal uses of computers.

It is well known that numerous concepts for humanoid robots were implemented throughout the Ancient Greek era. Daedalus, who is thought to have governed the mythology of the wind, is one example of someone who attempted to build artificial humans. The goal of defining philosophers' systems of human mind has begun to be observed in history through the evolution of modern AI (Haenlein & Kaplan, 2019). The year 1884 is crucial for AI. On this day in history, Charles Babbage began working on a mechanical device that would display intelligent behavior. Claude Shannon first proposed the notion that a computer might play chess in 1950. Until the early 1960s, AI research progressed slowly. Journey of AI from yesteryears to the future can be seen in Figure 2.1.

Key Concepts and Definitions

Evolution of technology through the development of "Artificial Intelligence" ideas has created a larger range of possibilities for raising intelligence to infinite degrees (Abbasi et al., 2022). The world has undergone a revolution thanks to computers' capacity to continually learn (also known as machine learning [ML]) and create a wide variety of intelligence notions. "Artificial Intelligence" refers to the application of algorithms and ongoing literacy appear to have mortal problem-solving and decision-making skills utilizing computers and other devices.

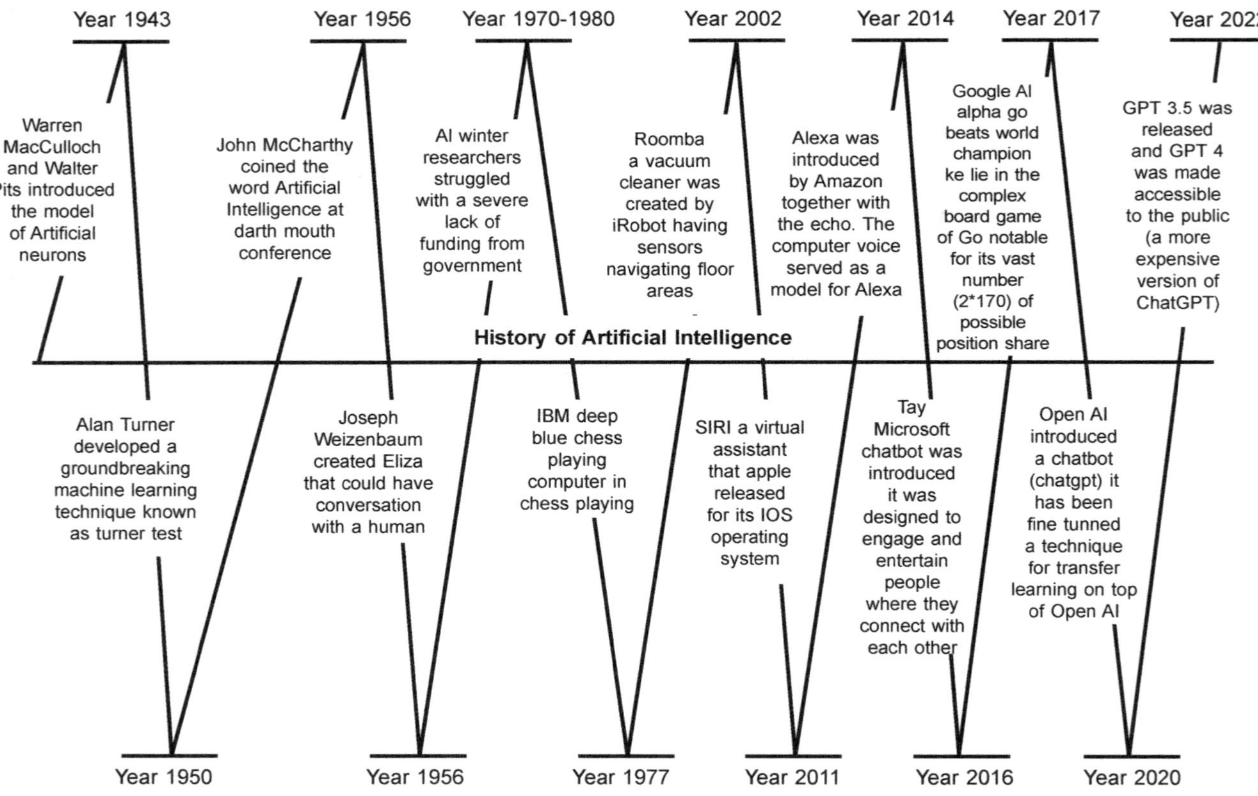

Figure 2.1: AI's development across time.

Introduction to Artificial Intelligence (AI)

Although there are other definitions of AI that have appeared in recent decades, John McCarthy provides the following one: "Building intelligent machines, particularly clever computer programs, is a scientific and engineering endeavor". AI shouldn't be restricted to methods that can be seen by biological means, even if it is connected to the related task of using computers to understand human cognition (Li & Du, 2017).

However, decades ago this term began to exist. Alan Turing's groundbreaking book, *Technology and Intelligence in Computing*, which was released in 1950, served as a marker for the beginning of the AI debate. Turing, "a pioneer in computer science", in this article, asks the following question: "Can machines think?" Additionally, he presents "The Turing Test", a procedure which includes a mortal questioning in an attempt to distinguish amongst a bone response that was written by a human and a textbook response that was created by a machine (Natale, 2021).

This test has received a lot of attention since it was published, yet despite this, it still represents a major contribution to the history of AI and is still a popular topic in the church since it uses linguistic generalizations. The release of *The Use of AI* by Stuart Russell and Peter Norvig is a Novel Technique in the following decade marked the beginning of one of the field's most influential workshops. One of the most significant works on the subject, *A Modern Approach*, was just released (Natale, 2021).

Important Definitions of AI from Reliable Sources

Merriam-Webster Definition for Artificial Intelligence

"A field of computational science dealing with automated modelling of smart actions." "A device's ability to imitate intellectual person behaviour" (Douthit et al., 2020).

Encyclopedia Britannica Definition of AI

"AI is defined as the capacity of a digital computer or computer-controlled robot carry out activities frequently linked with smart individuals." Intellectual organisms are capable of adapting to their surroundings.

English Oxford Digital Dictionary Meaning of Artificial Intelligence (AI)

"Theorising and developing computer systems capable of carrying out tasks typically demanding intellect from humans, such as seeing, hearing, making choices, and translating words into other languages."

Modern Dictionaries Definition of AI

As a subfield in the science of computers and emphasize the way robots may replicate human intellect rather than completely acquire cognitive abilities.

Typically, people fund the development of AI to achieve one of these three goals:

- Create systems with "strong AI" that fully mimic human thought.
- Without understanding how human reasoning functions, systems should just function.
- Apply human logic to situations as a human, but do not create it your ultimate objective.

How Vital is Artificial Intelligence?

The potential for AI to alter how we live, work, and play makes it significant. Automation of human jobs including customer service, lead creation, fraud detection, and quality control have been successfully applied in business. AI is capable of several tasks considerably more effectively than humans. ML algorithms usually complete tasks quickly and with minimal errors, particularly when it pertains to routine tasks like evaluating a large number of legal documents to ensure essential fields are accurately filled in. AI can provide businesses with operational insights they may not have known about due to the enormous data sets it can process. Product design, marketing, and education will all benefit from the fast-growing community of generative AI technologies.

In fact, improvements in AI methods have given rise to totally new business options for some larger corporations in addition to helping fuel an explosion in efficiency. It would have been difficult to conceive employing computer software to connect passengers with taxis before the current wave of AI, but Uber has achieved Fortune 500 status by doing precisely that.

Alphabet, Apple, Microsoft, and Meta are just a few of the biggest and most prosperous businesses in existence today. These businesses leverage AI technologies to streamline operations and outperform rivals. For instance, AI lies at the heart of Alphabet subsidiary Google's search engine, Waymo's autonomous vehicles, and Google Brain, which developed the transformer neural network design that supports recent advances in natural language processing (NLP) (Artasanchez & Joshi, 2020).

Strong AI vs Weak AI

Weak AI, also known as Narrow AI or Artificial Narrow Intelligence (ANI), is AI which is being constructed given instructions to carry out specific duties. The overwhelming portion of AI nowadays is of poor quality. This type of AI is far from inept; it can do some amazingly complex functions, such as Siri from Apple, Alexa from Amazon, the Watson system from IBM, and tone-driving motorcars; maybe 'narrow' would be a better descriptor.

Artificial general intelligence (AGI) had been present to as general AI, is the capability of a computer to suppose like a mortal, have tone-apprehensive knowledge, and be suitable to learn, reason, and produce unborn plans. The term 'superintelligence', or "artificial super intelligence" (ASI), refers to a position of intelligence and capability beyond that of the mortal brain. Although strong AI is substantially theoretical at this point and has no practical operations, specialists in the subject are constantly probing its possibilities. Till then, works of science fiction such as *2001: A Space Odyssey's HAL*, the invincible, wicked computer assistant, may be the finest examples of ASI (Eban Escott, 2017).

Key Concepts of Artificial Intelligence (AI)

The topic of AI is wide and expanding quickly. Deeper comprehension of AI's potential and uses can be attained by looking into these ideas in more detail and keeping up with recent research and advancements. The capabilities and uses of AI have been greatly broadened by recent developments in the field. Below are a few key concepts of AI.

Machine Learning (ML)

Automated learning pertains to a computational intelligence field that allows computers to learn from experience and progress without explicit programming. It has grown in elevation lately as a result of the multitudinous practical operations it has in a variety of diligence. Here, we'll look at the basics of machine literacy, go into more delicate motifs, and bandy how it's being used to break real-world problems.

The field of AI encompasses it. It tries to make robots more like humans in their conduct and decision-making by giving them the capability to learn and design their own programs. On the basis of the machines' experiences during the process, the method of learning has been computerized and enhanced. The computers are supplied high-quality data, and multiple strategies are used to generate models based on ML in order to educate the machines on this data. The kind of data at hand and the kind of action that needs to be automated will determine which algorithm is used (Ed Burns, 2023). Comprehend the example given in Figure 2.2.

Deep Learning (DL)

The term "Deep Learning" (DL) was founded on the concept of a neural network made up of computers. DL gained importance as a result, restarting artificial neural network research, thus the name "new-generation neural networks" in some context. Which is due to deep networks' great performance in a range of classification and regression problems when they are properly trained (Mohammad Mustafa Taye, 2023).

The information in everyday life can take numerous forms, nonetheless for models based on DL, it is typically reasonable to express it as follows:

- *Logical Data:* It is any sort of data in which its sequence is important, or an accumulation of patterns. The model needs to explicitly recognize the

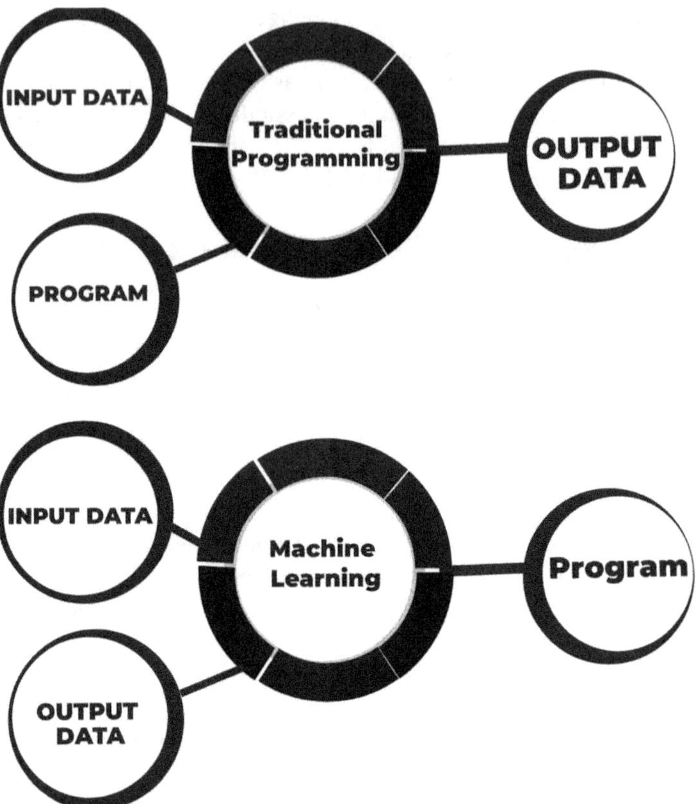

Figure 2.2: Machine learning vs traditional programming.

sequential nature of the incoming data when creating it. Text streams, audio clips, clips of video, and time-varying data are examples of logical data.

- *2D Data or an Image:* A matrix is the building block of a digital picture. It is a rectangle-shaped collection of numbers, symbols, or expressions organized in rows and columns in a 2D array of numbers. A digital image's four essential elements or qualities are matrix, pixels, voxels, and bit depth.
- *Tabular Information:* A tabulated information sets primary elements are rows and columns, respectively. As a consequence, tabulated information sets include data that is structured into columns in the same way that the information in a database table is. Every entry (column) must be named, and each field may only contain data of the specified kind. In general, the data is logically and systematically sorted into rows and columns depending on the data's features or aspects. Using DL models that can learn successfully from tabular data, we can develop intelligent systems that are driven by data (Sarker, 2021).

Natural Language Processing (NLP)

Theoretical linguist Noah Chomsky held a unique position since he changed the study of syntax. Chomsky was one among the earliest twelfth-century engineers to come up with structural ideas (Chomsky, 2009). It is basically classified into two categories: The first level, which corresponds to natural language, and second level, that incorporates speech recognition technology. Co-reference resolution, discourse analyses, machine translation, morphological segmentation, named entity identification, optical character recognition, part of speech tagging, automatic summarization, and other processes are a few of the NLP tasks that have been studied. Among other tasks, automated translation, named entity identification, and detection of optical characters have direct applications in everyday life. Automatic summarizing creates a clear descriptions of many different kinds of texts are provided or in-depth details of texts of a particular category (Khurana et al., 2023).

NLP is a methodical strategy used by computers to learn how people use, apply, and comprehend language. With the aid of AI, the methodologies and processes by computers have evolved to the point where they can understand and alter text. The application of advanced NLP in industry is being benefited by Industry 4.0. The term "fourth industrial revolution" refers to a new level of innovation, technology, and modern scientific notions of growth in the economy along with elevated quality of life (Mah et al., 2022).

Reinforcement Learning (RL)

RL which stands for reinforcement learning, a subfield of AI, is concerned with creating models and algorithms that let agents learn from their surroundings and make decisions. To maximize cumulative rewards over time, an agent must interact with its environment, learn from feedback (rewards or penalties), and then optimize its actions in the light of that knowledge. In a variety of fields and applications, RL is essential in the current world (Bhatt, 2018). Here are a few illustrations:

1. *Robotics:* RL is used in robotics to help machines learn difficult jobs and control their surroundings. Through the use of RL approaches, robots can learn how to grip items, maneuver around obstacles, and execute complex movements, increasing their adaptability and effectiveness.
2. *Game Playing:* RL has had a lot of success in the gaming industry. In games like chess, go, and poker, it has been crucial in training agents that have outperformed humans in terms of performance. Real-time learning (RtL) algorithms build methods to beat adversaries through iterative testing and optimization.
3. *Resource Management:* The allocation and management of resources are optimized using RL approaches. For instance, RL can learn to regulate power generation and distribution in energy systems to reduce costs or

increase efficiency. Additionally, RL can improve scheduling and routing in transportation systems, resulting in better logistics and less traffic (T. Zhang & Mo, 2021).

Knowledge Representation and Reasoning

A key component of AI is knowledge representation and reasoning (KR&K) that focuses on the illustration and application of knowledge to tackle challenging issues. It entails the creation of methods and formalizations to formally document and organize knowledge, as well as design of reasoning processes to provide insightful conclusions and wise choices based on that knowledge. It acts as the foundation for creating intelligent systems that can comprehend complex information and reason about it. Here are some main justifications for its importance:

1. *Effective Problem Solving:* KR&R helps AI systems to successfully address complex problems by describing knowledge in a structured and organized manner. The system can use it to store and retrieve pertinent knowledge, analyze it, and come up with intelligent answers.
2. *Learning and Adaptability*: Knowledge representation that is effective enables AI systems to pick up new information and skills from existing data. It enables systems to learn new information, update old information, and modify their thinking and decisions in response to new data or altering situations.
3. *Domain Knowledge and Problem Solving:* KR&R makes it possible to gather and use domain-specific knowledge. It lets specialists in a field to incorporate their expertise and perceptions into AI systems, enabling them to carry out difficult tasks and offer insightful information in a particular field (Richard Fikes, 2020).

Type of AI and Their Applications

Categories of AI

Two primary classifications are based on the capabilities and the functionalities of the technology. The flowchart explains the many categories of AI (Figure 2.3).

AI Based on Capabilities

Narrow AI

Narrow AI, commonly is limited to performing a single narrow task. It develops along a specific group of cognitive abilities. Narrow AI applications are becoming increasingly prevalent in our daily lives as ML and DL techniques progress.

Due to its limited training, narrow AI cannot execute tasks that are outside of its purview or capabilities. It is therefore frequently referred to as "weak AI".

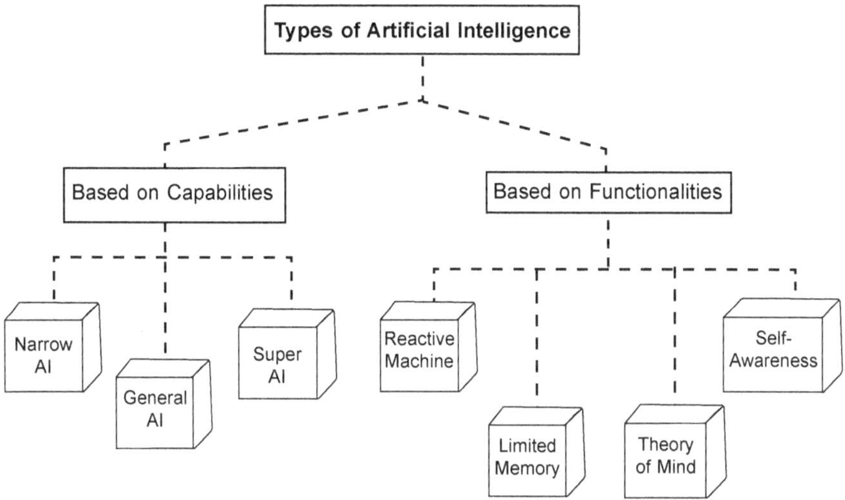

Figure 2.3: Types of AI according to their skills and functionalities.

Examples of Narrow AI

- *Voice Assistants:* Voice assistants like Siri, Alexa, and Google Assistant are created to listen to voice requests and carry out specified actions like playing music, creating reminders, and managing smart home devices. Within their prescribed domains, they excel at understanding voice inputs and processing spoken language.
- *Image Recognition Systems*: Image recognition systems examine and recognize objects, patterns, or features inside images using AI algorithms. These algorithms are frequently utilized in applications like object identification in self-driving cars, facial recognition to unlock smartphones, and picking out specific objects in image searches.
- *Virtual Personal Assistants:* AI-powered applications or programs that support consumers with various activities known as virtual personal assistants. According to user requests, they can arrange meetings, make reservations, send messages, or give information. Examples include Samsung's Bixby, Apple's Bixby, and Cortana from Microsoft.
- *Systems for Detecting Fraud:* These systems examine massive amounts of data and look for patterns that point to fraudulent activity. Fiscal associations use these technologies to identify and stop fraudulent deals, credit card fraud, and identity theft.
- *Language Translation*: To remotely convert spoken words or texts from one language to another, language translation systems make use of AI techniques, specifically ML and neural networks. With the help of tools like Google Translate, these systems have undergone significant upgrades over time and are now providing translations that are fairly accurate (Krupansky, 2017).

General AI

General AI, referred to as Strong AI, is capable of learning and comprehending everything a person is capable of doing in terms of cognitive endeavor. It enables a computer to apply information and skills in multiple contexts. Strong AI has not yet been developed by AI researchers. They would have to devise a way to program computers with the complete range of cognitive abilities necessary for consciousness.

Examples of General AI

Although actual General AI does not yet exist, the following hypothetical scenarios show some of its possible capabilities:

- *Human-Level Robotics:* General AI in robotics would entail building humanoid robots that are capable of carrying out human-level tasks. Advanced physical abilities such as dexterity, mobility, and the capacity to traverse and interact with the physical world would be present in these robots. They would be able to comprehend and react to complicated instructions, gain knowledge from mistakes, and adjust to novel circumstances.
- *Complex Problem Solving*: General AI may be used to break complicated issues in a variety of disciplines, including engineering, drug, and scientific exploration. It would be suitable to comprehend the craft of a situation, examine enormous quantities of data, and put forth new answers or propositions. This might hasten developments in fields like medicine discovery, climate modeling, or complex system optimization.
- *Language Understanding and Reasoning*: The intricacies, context, and idiomatic expressions of human languages would be deeply understood by a General AI system. It would be capable of having logical discussions on a variety of subjects, understanding and interpreting difficult materials, and having natural and intelligent dialogues.
- *Self-improvement and Learning*: A General AI would be able to continually gain knowledge and improve itself. Over the course of time, it would be able to learn fresh knowledge, enhance its own efficacy, and refine its present models. Because of this, it would be capable of establishing novel abilities, expanding its knowledge, and responding to changing circumstances.

Super AI

Super AI has intelligence above human limits and is adept at managing any assignment more than a person. In the theory of artificial superintelligence, AI has progressed to the point where it is so similar and so sensitive to human experiences and feelings that it not only understands them but also generates its own sentiments, desires, convictions, and objectives (Müller & Bostrom, 2016). Currently, there is no evidence to support its continued existence. Allowing, mystification-working, independent judgment, and decision-timber are some of the pivotal traits of super AI.

The development of generic AI devices is presently the main focus of researchers from all around the world. Building such systems will take a lot of time and money because general AI systems are still a relatively new field of study.

Examples of Super AI

Super AI, commonly referred to as Artificial General Intelligence (AGI) that exceeds human-level intelligence, is completely conceptual because it goes beyond what AI is at present capable of. Here are some speculative situations to show the potential extent of super AI, but it is currently only a concept:

- *Scientific Discovery and Breakthroughs:* Super AI's ability to analyze enormous volumes of data, spot patterns, and draw links between different fields of study could speed up scientific research. By creating hypotheses, running virtual experiments, and providing fresh perspectives, it may help to advance fields like medicine, materials science, or fundamental physics.
- *Advanced Spaced Exploration:* Super AI might be essential to future space exploration efforts. With little assistance from humans, it could interpret astronomical data, devise the best trajectories, and control complex robotic systems. Super AI might help in the search for new celestial objects, the solving of cosmic mysteries, and the execution of sophisticated space missions that are currently out of our reach.
- *Extensive Data Analysis and Predictive Modeling:* Massive volumes of data from various places, such as social platforms, academic research, and the stock market, may be handled and examined by Super AI. It might discover unknown patterns, predict results thoroughly, and generate insightful knowledge that would be beneficial for decision-making in a wide range of professions, including financial services, medical care, and the field of social science.
- *Advanced Autonomous Systems:* Super AI may result in the creation of extremely sophisticated autonomous systems that can carry out difficult tasks in changing situations. This might include robots that can do complex surgeries with utmost precision or autonomous cars with unmatched safety, efficiency, and adaptive capabilities (Kanade, 2022).

AI Based on Functionalities

Reactive Machine

A reactive machine the most common type of AI that does not keep memories or base decisions on past experiences. It only functions with current data. They take in the environment and respond to it. Reactive machines are given certain duties to complete, but they lack capabilities outside of those tasks.

Example of Reactive Machines

Automated traffic signal system is an illustration of a reactive machine. To regulate the timing and sequencing of traffic lights at junctions, these systems are made to respond in real-time to inputs like traffic flow and sensor data. The technology immediately decides when to change the lights to improve traffic flow and ensure safety after analyzing the current traffic circumstances (Ali et al., 2021). Reactive traffic signal systems lack memory and the capacity to take into account past data or forecasts. They merely respond to the inputs that are now there to decide on the proper signal alterations. They are crucial for effectively managing traffic congestion, cutting down on delays, and enhancing all aspects of traffic management in metropolitan settings.

Limited Memory

Small Memory AI learns to make decisions using historical data. Such systems have temporary memory. They are permitted to use this historical information for a limited time but are not permitted to add it to a library of their experiences. Autonomous vehicles employ this kind of technology.

Examples of Limited Memory

- *Chatbots*: During a conversation, chatbots frequently rely on a small amount of memory. They keep a short-term memory of the most recent communications with the user, enabling them to respond appropriately and keep the dialogue flowing.
- *Recommendation Systems:* To keep track of users' most recent activity and preferences, recommendation systems in e-commerce or streaming platforms frequently utilize limited memory. The accuracy and relevancy of the suggestions are increased by using this data to create customized recommendations based on the user's most recent interactions.
- *Fraud Detection Systems:* In banking or online transactions, fraud detection systems use a finite amount of memory to examine recent transaction patterns and spot possible fraud. These systems can identify irregularities and take the necessary steps to stop fraudulent transactions by taking into account a short-term memory of previous transactions.

Theory of Mind

Concept of mind AI is a cutting-edge class of technology that only exists in concept. A deep grasp of how the people and things in an environment can change feelings and behavior is necessary for this form of AI. It ought to be able to comprehend the feelings, opinions, and thoughts of others. Even though this sector has seen significant advancements, this particular AI still needs work.

Example

Real Life Example: Kismet is one application of the theory of mind AI. Kismet, a robot head, was developed by a Massachusetts Institute of Technology (MIT) researcher in the late 1990s. Kismet has the capacity to mimic and recognize human emotions. Despite the fact that Kismet can follow eyes and signal attentiveness to individuals, these two capabilities are still major advances in theory of mind AI.

Self-awareness

Self-awareness is just speculation. These systems can detect human emotions and comprehend their internal characteristics, circumstances, and moods. These machines will be more knowledgeable than the human mind. This kind of AI will not only have the capability to identify and spark feelings in people with whom it communicates, but will also enjoy its own feelings, requirements, and beliefs.

The use of AI in daily life is now best illustrated by Apple's Siri, Google Now, Amazon's Alexa, and Microsoft's Cortana. Users of these digital assistants can use them to check their calendars, perform online searches, and provide instructions to another program, among other things.

Applications of Artificial Intelligence

AI has a wide range of operations across colorful diligence and disciplines. Some crucial operations of AI include:

AI Shopping App that Personalizes Online Purchases

- *Customized Retail:* The use of AI technology enables recommendation engines, which improves how you communicate with your customers. These recommendations take into account the user's browsing history, preferences, and interests. It enhances brand loyalty and aids in the strengthening of relationships with your customers.
- *AI-Powered Helpers:* For online shoppers, chatbots and virtual assistants improve the customer experience. NLP is used in the conversation to make it sound as personal and natural as possible. These helpers can also communicate in real time with your customers.
- *Scam Avoidance:* Credit card theft and fake reviews are two of the most significant issues that e-commerce companies face. By considering use trends, AI can help to reduce the likelihood of credit card theft. After reading user evaluations, many shoppers decide whether or not to buy a product or service. The use of AI can help identify and manage fake reviews.

Education-related Uses of AI

Humans have the largest impact in the education sector, but AI has also begun to gain ground there. Even in the sector of education, the progressive adoption of AI has improved faculty productivity and allowed them to put more of their attention on the students rather than on office or administrative work.

- *Automated Administrative Tasks to Support Teachers:* AI can help educators with both task-related and non-task-related responsibilities, such as facilitating and automating personalized messages to students, back-office tasks like grading paperwork, planning and facilitating parent and guardian interactions, and routine issue feedback facilitation. These tasks include managing enrollment, courses, and HR-related issues, as well as enrollment, course, and HR-related matters.
- *Creating Intelligent Content:* Information like video lectures, conferences, and textbook guides may be digitized using AI. We may adapt learning materials and use different interfaces, such animations, for students in a range of grades. AI helps to provide a rich learning experience by creating and providing audio and video summaries and thorough lesson plans.
- *Voice-activated Technology:* It enables students to access extra learning resources or assistance even when a lecturer or teacher is not actively involved in their instruction. Making temporary handbooks at a lesser cost and making it easier to offer answers to commonly asked questions are both benefits of this.

Lifestyle Uses for Artificial Intelligence

The way we communicate is significantly impacted by the usage of AI. Many of those topics will be covered.

- *Autonomous Vehicle Operation:* Automakers like Toyota, Audi, Volvo, and Tesla employ ML to train computers to think and grow like people when it comes to driving in any environment, and object recognition to minimize accidents.
- *Recognition by Facial Features:* Our preferred devices, such as phones, laptops, and PCs, employ recognition by facial feature algorithms to detect and identify people in order to give safe access. Outside of personal usage, facial recognition is a common use of AI in many industries, even in extremely secure environments.
- *Recommendation Framework:* Many of the platforms we use every day, such as e-commerce, entertainment websites, social networking, video sharing platforms like YouTube, etc., utilize the recommendation Framework to gather user data and provide tailored suggestions to consumers to increase engagement. This widely used AI technology is used in almost all sectors.

AI Uses in Robotics

Robotics is another field in which AI is regularly used. Real-time updates are used by AI-powered robots to identify obstacles in their way and instantly plan their paths around them. It is suitable for:

- transporting commodities in warehouses, factories, and hospitals
- cleaning huge equipment and offices
- inventory control.

Healthcare Applications of AI

In the healthcare sector, AI is used extensively. AI applications are used in the construction of cutting-edge technologies that can recognize cancer cells and diagnose disorders. AI can help in the analysis of chronic diseases with lab and other medical data to guarantee early identification. To discover novel treatments, AI combines historical data with medical understanding (Adam Bohr, 2020).

AI Applications in Chatbots

AI chatbots can interpret natural language and respond to users of the "live chat" feature that many companies provide for online customer assistance. AI chatbots are useful when combined with ML and may be integrated into a wide range of websites and apps. AI chatbots will eventually be able to build their own databases of responses in addition to using information from integrated, pre-existing answers. As AI develops, these chatbots may effectively manage customer concerns, respond to simple inquiries, improve customer service, and provide round-the-clock help. Overall, these AI chatbots can help increase customer satisfaction.

AI in Automotive Industry

By incorporating AI into the production process, automakers can take advantage of smarter factories that increase efficiency and reduce costs. AI can be applied to designing automobiles, supply chain optimization, using robots on the factory floor, using sensors to improve performance, auto assembly, and post-production tasks.

In 2021 and 2022, supply chain issues and disruptions plagued the automotive industry. AI can help with this as well. By anticipating and replacing supply chains as necessary, AI enables businesses to identify the challenges they may face in the future. Routing issues, volume estimates, and other issues could be helped by AI.

Artificial Intelligence (AI) in Architecture and Algorithms

One of the most well-liked academic disciplines in the twenty-first century is AI. Although it may have scholarly origins in ancient history, it didn't really enter the global scene until the last decade of the twentieth century. AI is approached by diverse fields including engineering, medicine, marketing, thriftiness, etc. as a bolstering and influencing force for their creative work. It seems obvious that AI will eventually become a part of daily life, but it is crucial to consider what role it will play and how AI systems and human society will interact (Smith, 2018).

The Impact of AI in Architectural Design

The design of every built environment is the responsibility of the discipline of architecture. It offers a harmonious combination of interdependent components, necessitating the use of all resources, including technology, to advance itself. Technology has transformed design into a scriptive practice in which blueprints and models are utilized to foretell the future and where every detail must be figured out before construction can start (Ludewig, 2003). It also aids in achieving complex design goals and forms. Architects make use of the potential provided by AI to ensure that (Bhatt, 2018). As a result, it is one of the fields that could reinvent itself thanks to advances in AI. Consequently, the following inquiries arise: How might structure enable data and algorithms useful instruments to advance the industry? What methods may be used to use AI in design effectively? How could AI become an essential part of design?

Some present and prospective technologies, such as BIM software and building physics analysis tools, may help to partially answer the problems raised above, and parametric design interfaces can help to partially address these concerns. In addition to this, and perhaps more crucially, a fundamental understanding that views to ensure the long-term survival of AI in the field of architecture and a strong coordination between them, AI as a basic and natural component of architectural education must be provided. The main topic of discussion in architecture has been the direction of architectural education with the recent increase in computer-based applications (Guney, 2015).

The Gathering and Processing of Data

Data processing is the core of AI, and it is presently happening at a very high volume and speed. As a result, the practice of architecture benefits from this evolution. The architects must process a lot of information during the initial design phases, including legal requirements, analyses of the physical surroundings, user requirements, operational specifications, historical examples, etc. They are all interested in processing information (Cui & Zhang, 2021). Without the use of computer technologies, analyzing all this data might not be feasible, but AI offers significant assistance. Additionally, speeding up the indexing and categorization of data needed for the initial design phases has an unseen but significant impact on the design process (Deterding & Waters, 2018).

AI to Generate Design Possibilities

Commercial AI-based architecture software has already matured to a high level. While some of the products are standalone programs, others can be used like plugins, additional sophisticated programs. Among them, called Dynamo, is a Revit plugin which allows consumers to evaluate data and develop original algorithms using virtual programming. It allows users to automate procedures using a graphic algorithm editor's logic (Saad et al., 2020). The algorithmic design software CATIA, is an acronym for Computer Aided Three-Dimensional Interactive Application and was developed in 1977. Throughout the world, renowned architectural offices utilize the software. The graphical parametric is another form-generating software Grasshopper which is a subordinate of

Rhinoceros. It is utilized in many design scales and facets. Schneider and his colleagues, for instance, used Grasshopper to create an urban design proposal during a classroom exercise.

Building Information Modeling (BIM)

In the field of modern architecture, one of the most often used phrases is building information modeling (BIM). The multidisciplinary approach of BIM, which strengthens ties between various building industry players including architects, engineers, and contractors, is another ground-breaking innovation. BIM involves a process shift as well as a technical change. BIM's primary goals are to reduce project costs, increase production and quality, and expedite project completion (X. Zhang et al., 2018). BIM refers to a virtual environment simulation of the building design. The finished model, called a BIM, is an information-rich, object-oriented, intelligent, and parametric digital representation of the facility from which views and data appropriate for various users' needs can be extracted and analyzed to produce information that can be used to make decisions and enhance the process of delivering the facility (X. Zhang et al., 2018). Because of the BIM software's multilayered, multidisciplinary structure's tremendous complexity, AI must serve as its foundation.

Analysis of Building Efficiency

Building efficiency evaluation is the process that looks at how a building behaves under particular circumstances. Although it is not a new concept, AI and BIM have made it more manageable, dependable, and reproducible. For more efficient and effective outcomes, performance analysis software or plugins are now employed in conjunction with BIM software through interoperable methods. Building performance analysis can be divided into two basic categories: structural performance and energy performance. Both have value in their own unique ways. The following chapters of the essay discuss the two sorts of analysis that AI has made available.

Energy Performance Analysis (EPA)

Building façades, for example, might have multiple emphasis points for energy performance study, such as visual and thermal comfort. Software for EPA is assessed using the following criteria: the tool's precision and ability to replicate complex and finely detailed architectural elements. The interface's functionality as well as data handling. Incorporation of knowledge provides a foundation for innovative design. Many EPA tools are currently available, including Ladybird, Honeybee, Geco, and Heliotrope-Solar. They generally function under other applications as plugins. The Ladybird tool, for instance, will work with Grasshopper to provide a single parametric platform to cover the whole spectrum of environmental studies.

Robotics in Design of Architecture

People judge and worry about things they don't comprehend. We must overcome several barriers, including a lack of data, operation, support, collaboration, and other elements. However, armature needs to adapt and develop just like any other assiduity. As decisions are made, new information becomes increasingly important. These barriers can be jumped by AI. It may develop judgments, extract data, and analyze that data based on the information gathered. The felicity, quality, and variability are crucial while creating a design. Engineers are now operating privately. They physically define and prepare places, optimize the design, and provide quality work.

However, it does take some time. The entire procedure necessitates a lot of labor and research. Using computerized architectural design, architects may use a process that is more objective and scientific. AI is meant to improve an architect's job, not to replace it.

Using Algorithms to Improve Design

Better buildings are being made by algorithms. Algorithms can assist in the refinement, informing, and creation of fresh designs made possible by CAD (computer-aided design), text-to-image AI algorithms as Dall E 2 and Mid Journey, and Stable Diffusion, among others. AI allows architects to work with more unusual shapes.

AI algorithms will be used to create augmented reality models of buildings, allowing constructors and architects to take risks with design ideas and make informed decisions in a controlled environment.

Intelligent Design Systems

There are various businesses that use AI. It has the ability to gather enormous amounts of data and use that data to forecast the future. For example, businesses already use it to enhance language processing, streamline processes, complete repetitive activities, and so on.

The goal of architecture is to use creativity to maximize available space, it has the potential to fundamentally transform this industry. In order to redesign cities, it can aid architects in understanding traffic volumes and patterns. Although it may appear to be a difficult task, AI can make it easier and supply all the potential data points.

Automated and Smart Architectural Design

AI algorithms will be used to automate the construction process, producing potential designs that satisfy established standards and assisting designers in making better informed decisions. Digital assistants powered by AI will be used to manage building activities, providing information and assistance to inhabitants while also boosting their overall experience.

Maintenance Planning

Building maintenance personnel will be able to solve problems before they arise by using AI algorithms to predict equipment breakdowns. This will save downtime and enhance building efficiency. Algorithms based on AI will be used to increase structure effectiveness, reducing energy consumption and the carbon impact of a structure.

Building Automation Maintenance

Building maintenance duties like air quality monitoring and equipment failure detection can be automated using AI algorithms. This may reduce interruptions, improve the efficiency of construction, and free up additional time for designers and constructors to focus on other aspects of the design process. AI algorithms may be used to identify opportunities to reduce construction expenses, such as through the advancement of building materials and methods (Saad, Alaloul and Ammad, 2022).

AI will Decrease the Time it Takes to Create a Correctional Facility

The conception to figure delay in framework can be implicitly reduced by AI. Manufacturers can quickly generate and evaluate many design choices by using AI algorithms, which will ultimately help them develop the final product with greater efficiency.

Generative engineering software is one method that AI may be applied to armatures. Using algorithms, this kind of software generates a range of design possibilities based on a given set of input data, such as the structure's size and location. Additionally, engineers may quickly assess these choices and select the bone that best fits their needs.

BIM applications is another way that AI may be included into construction. Engineers may create digital versions of their layouts using BIM applications, and these models can be dissected applying AI algorithms to find hidden problems and offer solutions. This can assist engineers in identifying and fixing issues early on, minimizing delays and costs related to making modifications later in the development procedure.

AI may also be utilized to improve a building's energy efficiency. Designers can assess several potential designs and get the ideal outcome to reduce energy usage by using technological expertise and computation (Farzaneh et al., 2021).

Artificial Intelligence (AI) in Material Science and Construction

Material science and building have been transformed by AI, which has made sophisticated computational tools and automation capabilities available. In the present world, AI is fundamentally changing these sectors, revolutionizing material discovery, design optimization, building efficiency, and maintenance.

Artificial Intelligence and Material Science Integration

Over millions of years of human history have been significantly defined by accessories, from the Paleolithic Age through the upcoming fourth artificial revolution. Exploring the connections between physical structure, process, components, and function is a key component of material knowledge. The advancement of mortal civilization will be aided less by the emergence of novel accessories. The area of accessories knowledge has developed over several centuries, and a significant amount of data has been gathered.

The growth of material knowledge has a fresh beginning with the rise of AI. After more than 60 years of evolution from the straightforward perceptron to the intricate multidimensional neural networks, AI has established a fundamental algorithm framework and a crucial tackle basis. In fact, a highly developed AI system trounced world champions in a variety of competitions, including chess, go, quiz games, and various other sports (Ray, 2018). The scientific community of practical expertise has taken notice of AI's superior mining of information capabilities. The Turing Award laureate Jim Grey presented "the fourth model of knowledge" in 2007 at the NRC-CSTB meeting. It is a data-ferocious knowledge that uses big data and AI to condense a lot of supplied knowledge into unproven hypotheses that will serve as a roadmap for new scientific discoveries. Large-scale increase space or complex procedures, which are similar to some challenges in material exploration, can be handled by this approach. The interdisciplinary for equipment informatics, which combines tools wisdom and AI methods, can assist scientists in more effectively gaining the deceased connection among various variables, forecasting particular tools packages, directing the chemical amalgamation path, optimizing the procedure's criteria, and improving the human material description manner.

ML, an important branch of AI, which is currently experiencing enormous expansion, is the application of AI with the greatest promise in material science research. The ML foundations presented in the next part establish the framework for the discussion of research on material implementations of AI in the chapters to follow.

Basic Concepts of ML

ML refers to a computer's capacity to learn from a set of data and then identify the rules or information that underlie that data. To be more precise, ML is primarily broken down into four steps: data gathering, data visualization, method choice, and model improvement.

Data Collection

Data may be acquired through modeling, investigations, and through online records, and ML represents a type of information-driven program (Xia & Xu, 2021). Simulations are kin to molecular dynamics (MDs) and viscosity functional proposition (DFT) calculations. Physical components and structural details on various accessories are included in the data. Because of the limitations of the terrain and the experimental setup, there are a lot of missing, duplicated, and conflicting data in the field of accessories. Data cleansing is therefore reasonably essential to find and fix various offences in the original data. The average, minimum, or other statistical values of the characteristic are used, if appropriate, to fill in any missing data. The fundamental idea behind omitting records with repeated values that are indistinguishable from one another is sorting by trait values and including records with the same value. The linked algorithms include, amongst others, the first line approach and the sorted-neighborhood approach. Similar approaches have been used with perovskite data by integrating several things such as the Accoutrements Project Database and the Inorganic Crystal Structure Database. In the event of contradictory values, bespoke programs may be developed to assess if the data meet the criteria. These programs will depend on the allowable value spectrum and the broad significance of each factor. Data that falls outside the anticipated range or has incongruent features will be properly deleted. After the drawing is completed, data may be used for data visualization.

Data Representation

The process of converting unstructured data into representations suitable for algorithms is known as data representation. Even if the data we get is frequently mathematical, it could indicate that it is appropriate for the algorithm. When we want to solve mathematical issues, we need to include equations or plot-relevant graphics to aid in comprehension. The input data has to be in the right structure enabling ML algorithms to learn more efficiently. The model performs better the more suitable representation we employ.

One method of encoding structural information and physical characteristics is binary coding. Granda et al. (2018) suggested a robot for organic synthesis. By coding in binary the chemical input, the robot can study the reactivity of reagent combinations and anticipate unknown chemical reactions using a support vector machine (SVM) model.

Selection of an Algorithm

The most often used algorithms at the moment are artificial neural networks (ANNs), decision tree symbolic regression, and k-nearest neighbor (KNN). The sections that follow will give a quick overview of various techniques.

KNN is an efficient and straightforward classification and regression technique. Given a training dataset, the technique finds k items in a dataset that are most similar to the new datum, and the new datum will be classified in the category that appears most frequently. The selection of k, distance computation, and classification technique make up the algorithm. As the model complexity degree rises with k, the error in approximating the model will decrease but the estimated failure will rise. When evaluating two locations using a variety of distances, it can be achieved to obtain different results. Manhattan distance, Euclidean distance, and more alternatives are available through KNN. Since it removes the empirical error, KNN often uses majority voting as the classification rule.

Decision tree is one of the ML's most straightforward and effective algorithms. A decision tree is a representation of a classifier that receives input in the form of a number of attributes and values and generates a decision. Values for the input and output might be discrete or continuous. Boolean classification is used when the output only has two potential values and the inputs are discrete (Unler & Murat, 2010). The choice made by a decision tree is reached after a number of tests have been conducted. In a decision tree, each node is a test of the validity of an element from the information being provided, and the subsequent branches that arise from it are numerous potential values for the property. The amount that the method returns makes up each leaf node.

Symbolic regression is a traditional AI method, particularly genetic programming-based symbolic regression (GPSR). Due to the absence of the functional link between variables, it differs from conventional numerical regression. Instead, each putative function's functional form is acquired through chromosomal evolution. The chromosomes are made up of terminal bumps with variables and constants and a collection of interior bumps with symbols for fine operations. The evaluation function is the difference in error between the experimental data and the data that the function has fitted. The seeker with the smallest crimes and the topmost rigidity might produce seed first. Different chromosomes go through mutation and heredity, reiterate gradationally, and repeat this process until the stylish function and parameter set for a particular issue are discovered. For illustration, the magic angle in graphene, the density of normal hydrogen, and the hunt for descriptors of perovskite stability are all areas of material exploration where GPSR is applicable but where there's little previous information and an unclear link between the associated variables (Sha et al., 2020).

Model Optimization

A model can better fit training data with the aid of higher-degree polynomials, but if the degree is set too high, it may overfit the training data and underperform on the validation data. The two techniques for figuring out the degree of the polynomial are cross-validation and regularization, which directly reduce the weighted sum of the empirical loss and the model's complexity.

Loss function is frequently used to identify the model with the most manageable error rate. The goal of the loss function is to determine how much actual values differ from the expected values. By reducing the loss function, the best theory may be found. Cross-validation is only trustworthy when the samples used for training and validation are representative of the total population.

AI Applications in the Sciences of Materials

An increasing variety of sectors have adopted AI in recent years, and ML investigation in the discipline of material science is progressing swiftly, particularly since it can create novel materials and foretell diverse chemical syntheses. This section will examine how ML may be used to assist individuals to overcome obstacles in the design, synthesis, and processing of materials.

Boosted Simulation

The research technique for computational chemistry and the study of materials has reached its third wave. The phrase "structure-performance" describes a calculation that, in order to anticipate the performance of the materials based on the structure, principally use a regional optimization approach. The additional technique, called "crystal structure prediction", mainly makes use of an international optimization technique to estimate geometry and efficiency from component composition. The third phase, referred to as "statistically driven design", makes use of ML geometry to extrapolate components' compositions, structures, and properties, physical as well as chemical parameters. The search for high-performance materials has been hampered by the theory's shortcomings, and the model's parameters do not exactly match real-world circumstances like grain borders or transformed phases. In this regard, although later tests have shown that the conductivity of zirconium-doped lithium tantalum silicate will really be 10^{-5} S cm^{-1}, the DFT estimate of its conductance would be 10^{-3} S cm^{-1}. So, it's crucial to figure out how to apply ML to make up for simulation's shortcomings.

Atom2vec

Atom2Vec, an unsupervised ML methodology, only needed a few hours to reconstruct the structure of the periodic table of elements. Atom2Vec first learns to distinguish different atoms by looking at the list of compounds in the online database. Then, we use the simple principle of NLP: properties of a word may be deduced from words nearby, and chemical components are categorized according to their chemical surroundings. Additionally, a variety of ML models can use the vectorized atomic descriptor as an input since it offers a plethora of information about the periodic law of elements. This creates a potent new path for the numerical way to represent physical data in the future (Sha et al., 2020).

Raising the Scale of the Simulation

The corresponding energy or force field may be quickly identified once ML recognizes these recurring patterns since the theoretical calculation of the atomic force field contains some predictable repetitions. Since millions of atoms may move in a few nanoseconds as opposed to a few hundred in a few picoseconds, the simulation computation's duration and temporal range are much increased, and the results are improved. Chemical processes and intricate material structures, such corrosion and interfacial interactions, can be mimicked.

Because there are so numerous colorful infinitesimal surroundings and different kinds of bonds, creating secure interatomic capabilities in large-scale MD simulations of chemical processes at shells and interfaces is an extremely delicate task. Interatomic eventuality grounded on ANNs has just come into actuality, offering a fair approach for creating the implicit energy face of systems that are challenging to characterize using conventional eventuality. To confirm the perfection and validity of the interatomic eventuality of the ANN, Artrith et al. (2013) employed bobby and zinc oxide as reference systems. They also defined the CuZnO ternary combination system with oxide-supported bobby clusters. In general, the implicit energy of the NN is relatively accurate, with results that are numerous orders of magnitude more effective and near to the advised value of the abecedarian reference electronic structure. Due to the demand for a large number of training points, the product of NN implicit energy has lesser processing conditions compared to other implicit-energy computation styles. Still, it's clear that NN has benefits in large-scale operations where conventional electrical structure computations are challenging to break.

Bringing Down the Computational Cost

By using a standard simulation calculation, it's challenging to explore all implicit combinations in a respectable quantum of time due to the vast combination spaces of accoutrements. For example, the lowest known sulphide nanocluster Au15 (SR) 13 has a bimetallic arrangement that reaches 32000, making it extremely delicate to cut all possible configurations. Still, the computational complexity will be vastly dropped and the filtering speed will be boosted by numerous orders of magnitude if only a small portion of the data is used to train the ML model, and also the model is used to read the other combinations. To read the CO adsorption energy of nanoclusters, Panapitiya et al. (2018) suggested an ML model grounded on the stochastic timber fashion. First, the Ag-alloyed Au25 nanoclusters DFT simulation data training model was applied. The delicacy of the authors prognostications of the adsorption energy, which used a two-step point selection procedure and point engineering approach, was 0.78 (R2) and 0.17 (RMSE). The distribution of Ag tittles in Au25 had the topmost impact on CO adsorption spots, according to the scientists' interpretation of the arbitrary timber's main bumps. Other Au-grounded nanoclusters may be simply added to the ML model. The model is anticipated to be used as a webbing tool to identify particulars that are applicable for further thorough study.

AI Material Property Forecasting

The development of new products in a variety of areas may be significantly accelerated if the packing of accessories can be precisely forecast. ML techniques have so far been successful in predicting an extensive variety of substance characteristics, including electrical characteristics like band gaps, topological regions, and superconductivity in addition to mechanical characteristics like large quantities and shear moduli and thermal features like conductivity.

The amount of data available is a significant obstacle to AI's prediction of material properties, despite projects like the Materials Genome and the Materials Project. One method to get past this constraint is to use the transfer learning machine learning framework. According to a study by Yamada et al. (2019) "Transfer learning relies on the concept that various property types, such as physical, chemical, electronic, thermodynamic, and mechanical properties, are physically interrelated." AI can predict a characteristic with a high degree of accuracy by employing these "proxy properties", much like a human expert who can make observations based on a constrained amount of input data. Another study using transfer learning showed that AI is capable of making material quality predictions even in the lack of crystal structure knowledge.

Material Invention Using AI Assistance

The hunt for novel materials is perhaps the most intriguing use of AI in materials research since it has long been assumed that only humans are capable of discovering and designing new materials. Because there are so many possible results in the feature space, which includes all possible atom configurations along with the three spatial dimensions, this search is challenging. Despite recent advancements in 'classical' structure construction and selection algorithms as well as energy evaluation tools, this process still demands a lot of computer power (Pyzer-Knapp et al., 2022).

AI-assisted simulations can significantly speed up the rate at which we learn the energy of many configurations, accelerating the entire discovery process. The feature space surrounding a configuration must be exhaustively scanned to show that the configuration's energy is contained in a local optimum to verify that it is stable. However, given the amount of data involved, AI is the best technology to solve this issue. AI has already shown that it can reduce calculation time by 75% while assisting scientists in the discovery of hundreds of novel combinations.

But there are other ways to show a material's stability except through simulations. Because AI has the distinct capacity to derive insights from data, it can now make greater use of experimental data. In a recent study, researchers first expanded their use of unsupervised learning techniques, including autoencoder, to predict novel crystal structures by looking for patterns of similarity between element combinations given published crystal shape. This modern AI tool sets itself apart by being able to locate distinctive resources as opposed to those that are identical to those we currently have (Anton Zhang, 2021).

AI in Primary Phases of Construction

Gaining Business Advantage and Contracts Using AI and NLP to learn from previous projects and data, you may avoid starting from scratch on every project or pitch. AI may help increase a construction firm's success rate in the initial tender process by studying previous project bids, replicating the successful components, and avoiding the failure-causing variables. These learning algorithms might enhance a construction company's win rate, predict the likelihood of a go-or-no-go scenario, increase profitability, and boost project value (Ammad et al., 2021).

The Generation of Design Alternatives

Design may be aided by AI through the use of generative design, a process for creating forms that mimics nature's evolutionary method of creation. The system, like AutoDesk, is first given precisely defined design objectives, and then it investigates several variants of potential solutions to choose the optimal one.

Utilizing OCR (Optical Character Recognition) to Save Time

OCR technology may be used in the construction industry to swiftly search drawings and transform documents and photos into editable and searchable data, saving time compared to manually typing up or redrawing materials. Construction companies can have a vital amount of time and money by using OCR to scan drawings, name and number sheets, and associated documents in a variety of drawing programs (ElZahed & Marzouk, 2022).

AI During the Execution Phase of Construction

Construction Methodology

Engineers can receive advice from AI database systems on the optimal building approach for a site based on prior projects as well as current blueprints at the design stage. Engineers can now make critical decisions based on information that may not have previously been available to them (Saad, Alaloul, Ammad et al., 2022).

Administrative Functions

After construction has begun, AI may be utilized for administrative duties like letting employees enter sick days, holidays, and vacation days into a data system. The project will then be adjusted as necessary, and duties will be automatically delegated to other staff on the affected days.

AI in Structures

Both commercial and residential structures may include AI. Future improvements might lead to so-called cognitive buildings that analyze and upgrade themselves based on past data, even if they are now employed to optimize energy consumption, manage environmental aspects (such as lighting and temperature), and provide an extra layer of protection.

AI Enhancing Project Management in the Construction Industry

- *Drones:* By employing drones to take exact survey maps, aerial images of a building site, and monitor work remotely, an assignment may be finished more quickly and at a lower cost. Additionally, project managers may gain a fresh perspective on their job and be able to see potential issues that were not visible from the ground thanks to the aerial photographs.
- *Alerts:* Foremen can input job site activities or problems using field reporting software, which can be configured to generate an alert or message when particular keywords, such 'delay' or 'safety', are used. These notifications keep important parties informed about projects.

AI Enhancing the Safety and Profitability of Construction

- *Intelligent Assistants:* The notoriously high-risk, low-margin construction sector is in desperate need of an investment in technology that can boost efficiency. AI gives the construction industry the capacity to filter through countless data points, immediately and automatically classify them, and highlight crucial expenditure and risk information for management.
- *Safer and More Conforming*: AI tools, like Smartvid, may recognize workers in images and videos taken at a working site, determine whether or not they are wearing safety equipment, and notify safety officers of any violations. This can reduce danger and encourage good behavior.
- *Increased Output*: Eliminating data silos is one of AI's most practical uses in the construction industry. Spot-R, which enables team managers to view the current position of employees on their 2D drawings and 3D models, is an illustration of this. ML algorithms can measure productivity using this data and provide suggestions for enhancements.

Construction Industry Transformation Using AI

The construction sector is on the verge of a transformative upheaval, and AI is at its core. AI is revolutionizing the way we plan, build, and maintain buildings by streamlining operations, enhancing security, and promoting sustainability in previously unheard-of ways. Through the use of linked gadgets, sensors, and drones that collect real-time data, monitor progress, and manage resources, AI is converting construction sites into 'smart' sites. These technologies make it possible to manage projects more effectively and to take safer precautions and allocate resources more effectively (Alaloul et al., 2020). For instance, Caterpillar's Cat Connect system monitors equipment performance, fuel usage, and maintenance schedules using AI-powered sensors and data analytics, optimizing resource allocation and decreasing downtime.

Table 2.1: Major components of AI transforming construction industry

Machine learning	What it could do?	Challenge
Computers built to think like people, making and carrying out judgments based on large and varied amounts of data.	By generating predictions about things like optimizing cuts in steel beams in a structure that is under construction, this technology has the potential to decrease costs and waste.	Data from the construction sector includes numerous factors, such as the size and design of buildings or the environment at the jobsite. Making precise forecasts is difficult because of the larger number of variables (more dimensions).
Robotics	**What It Could Do?**	**Challenge**
Automated machines that carry out physical duties that are similar to those performed by humans are called robots.	Robots could play a bigger role in the construction sector, doing specialized, repetitive work, or operating in risky situations like the tops of towering structures. Engineers, general contractors, and craftsmen may focus on activities that need higher skill levels if a robot was used on a project.	It might be difficult to use them on construction sites where there is much variety in the terrain, operational conditions, and design aspects. Robotics have a high initial cost as well as significant maintenance and repair costs.
Information-based systems	**What It Could Do?**	**Challenge**
The branch of Artificial Intelligence known as knowledge-grounded system; computer make judgment grounded on preliminarily acquired information.	The capacity to assemble, assay, and apply vast volumes of data from several sources in order to make intricate judgments. The knowledge base, which stores the data, and the conclusion machine, which analyzes the data in the knowledge base, are the two abecedarian factors of a knowledge-grounded system.	The data would originate from many sources and be of varying degrees of quality since the complexity of the construction industry involves several businesses, materials, people, industries and other entities on a given project. Another obstacle in this area is the propriety and legal issues around data sharing.

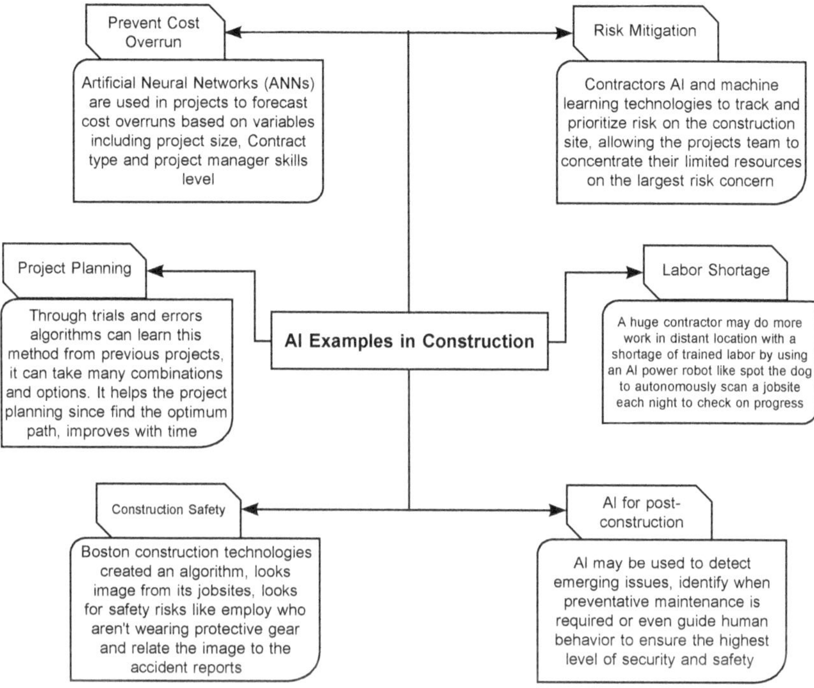

Figure 2.4: Examples of Artificial Intelligence applications in construction.

Examples of AI in Construction

AI and ML have a vast range of potential uses in the construction industry. In the business, information requests, unresolved problems, and modification orders are commonplace. ML can examine this sea of data like a knowledgeable assistant. This is how AI is already used in many applications as shown in Figure 2.4.

References

Abbasi, M.F., Bilal, M. and Rasheed, K. (2022). Role of Human Intuition in AI Aided Managerial Decision Making: A Review. *2022 International Conference on Decision Aid Sciences and Applications, DASA 2022*. https://doi.org/10.1109/DASA54658.2022.9765153.

Adam Bohr, K.M. (2020). *The Rise of Artificial Intelligence in Healthcare Applications*. Elsevier. https://doi: 10.1016/B978-0-12-818438-7.00002-2

Alaloul, W.S., Saad, S. and Qureshi, A.H. (2020). Construction Sector: IR 4.0 Applications BT. *In*: C.M. Hussain and P. Di Sia (Eds.), *Handbook of Smart Materials, Technologies, and Devices: Applications of Industry 4.0*, pp. 1–50. Springer International Publishing. https://doi.org/10.1007/978-3-030-58675-1_36-1.

Ali, A., Ud-Din, S., Saad, S., Ammad, S., Rasheed, K. and Ahmad, F. (2021). Artificial Neural Network Approach to Study the Effect of Driver Characteristics on Road Traffic Accidents. *2021 International Conference on Data Analytics for Business and Industry, ICDABI 2021*. https://doi.org/10.1109/ICDABI53623.2021.9655827.

Ammad, S., Saad, S., Bashir, M.T., Qureshi, A.H., Altaf, M. and Rasheed, K. (2021). Construction Accidents via Integrating Building Information Modelling (BIM) with Emerging Digital Technologies: A Review. *2021 3rd International Sustainability and Resilience Conference: Climate Change*. https://doi.org/10.1109/IEEECONF53624.2021.9668066.

Anton Zhang. (2021, November). *How Does AI Reshape Material Science*. Paper published in Rice Ken Kennedy Institute.

Artasanchez, A. and Joshi, P. (2020). *Artificial Intelligence with Python: Your Complete Guide to Building Intelligent Apps using Python 3.x*. Packt Publishing Ltd.

Artrith, N., Hiller, B. and Behler, J. (2013). Neural network potentials for metals and oxides – First applications to copper clusters at zinc oxide. *Physica Status Solidi (B)*, 250(6), 1191–1203. https://doi.org/https://doi.org/10.1002/pssb.201248370

Bhatt, S. (2018). *Reinforcement Learning 101*. Published in Towards Data Science.

Chomsky, N. (2009). *Cartesian Linguistics: A Chapter in the History of Rationalist Thought*. Cambridge University Press.

Cui, M. and Zhang, D.Y. (2021). Artificial intelligence and computational pathology. *Laboratory Investigation*, 101(4), 412–422.

Deterding, N.M. and Waters, M.C. (2018). Flexible coding of in-depth interviews: A twenty-first-century approach. *Sociological Methods & Research*, 50(2), 708–739. https://doi.org/10.1177/0049124118799377.

Douthit, B.J., Hu, X., Richesson, R.L., Kim, H. and Cary Jr, M.P. (2020). How artificial intelligence is transforming the future of nursing. *American Nurse Journal*, 29(10), 1513–1517.

Eban Escott. (2017). A guide to narrow, general, and super artificial intelligence. https://codebots.com> artificial intelligence

Ed Burns, G.L. (2023). *Artificial Intelligence (AI)*. https://www.techtarget.com/search enterpriseai/definition/AI-Artificial-Intelligence.

ElZahed, M. and Marzouk, M. (2022). Smart archiving of energy and petroleum projects utilizing big data analytics. *Automation in Construction*, 133, 104005. https://doi.org/https://doi.org/10.1016/j.autcon.2021.104005.

Farzaneh, H., Malehmirchegini, L., Bejan, A., Afolabi, T., Mulumba, A. and Daka, P.P. (2021). Artificial intelligence evolution in smart buildings for energy efficiency. *Applied Sciences*, 11(2). https://doi.org/10.3390/app11020763.

Granda, J.M., Donina, L., Dragone, V., Long, D.-L. and Cronin, L. (2018). Controlling an organic synthesis robot with machine learning to search for new reactivity. *Nature*, 559(7714), 377–381. https://doi.org/10.1038/s41586-018-0307-8.

Guney, D. (2015). The importance of computer-aided courses in architectural education. *Procedia—Social and Behavioral Sciences*, 176, 757–765. https://doi.org/https://doi.org/10.1016/j.sbspro.2015.01.537.

Haenlein, M. and Kaplan, A. (2019). A brief history of artificial intelligence: On the past, present, and future of artificial intelligence. *California Management Review*, 61(4), 5–14. https://doi.org/10.1177/0008125619864925.

Kanade, V. (2022). *What is Super Artificial Intelligence (AI)? Definition, Threats, and Trends*. https://www.spiceworks.com.

Khurana, D., Koli, A., Khatter, K. and Singh, S. (2023). Natural language processing: State of the art, current trends and challenges. *Multimedia Tools and Applications*, 82(3), 3713–3744. https://doi.org/10.1007/S11042-022-13428-4.

Krupansky, J. (2017). *What is an Intelligent Digital Assistant.* https://jackrupansky.medium.com.

Li, D. and Du, Y. (2017). *Artificial Intelligence with Uncertainty.* CRC Press.

Ludewig, J. (2003). Models in software engineering: An introduction. *Software and Systems Modeling*, 2(1), 5–14. https://doi.org/10.1007/s10270-003-0020-3.

Mah, P.M., Skalna, I. and Muzam, J. (2022). Natural language processing and artificial intelligence for enterprise management in the era of Industry 4.0. *Applied Sciences 2022*, 12(18), 9207. https://doi.org/10.3390/APP12189207.

Mohammad Mustafa Taye. (2023). *Understanding of Machine Learning with Deep Learning: Architectures, Workflow, Applications and Future Directions.* https://doi: 10.3390/computers12050091.

Müller, V.C. and Bostrom, N. (2016). Future progress in Artificial Intelligence: A survey of expert opinion. *In:* V.C. Müller (Ed.), *Fundamental Issues of Artificial Intelligence*, pp. 555–572. Springer International Publishing. https://doi.org/10.1007/978-3-319-26485-1_33.

Natale, S. (2021). *Deceitful Media: Artificial Intelligence and Social Life after the Turing Test.* Oxford University Press, USA.

Panapitiya, G., Avendaño-Franco, G., Ren, P., Wen, X., Li, Y. and Lewis, J.P. (2018). Machine-learning prediction of CO adsorption in thiolated, Ag-alloyed Au nanoclusters. *Journal of the American Chemical Society*, 140(50), 17508–17514. https://doi.org/10.1021/jacs.8b08800.

Pyzer-Knapp, E.O., Pitera, J.W., Staar, P.W.J., Takeda, S., Laino, T., Sanders, D.P., Sexton, J., Smith, J.R. and Curioni, A. (2022). Accelerating materials discovery using Artificial Intelligence, high performance computing and robotics. *NPJ Computational Materials*, 8(1), 84. https://doi.org/10.1038/s41524-022-00765-z.

Ray, S. (2018). *The Emergence of Artificial Intelligence.* https://towrdsdatascience.com>the-emergence-of-artificial-intelligence.

Richard Fikes, T.G. (2020). *Knowledge Representation and Reasoning—A History of DARPA Leadership.* Wiley Online Library.

Saad, S., Alaloul, W.S. and Ammad, S. (2022). Role of cyber-physical systems in smart cities. *In:* D.W.S. Alaloul (Ed.), *Cyber-Physical Systems in the Construction Sector* (1st Edn.), p. 19. CRC Press. https://doi.org/9781003190134.

Saad, S., Alaloul, W.S., Ammad, S. and Qureshi, A.H. (2022). A qualitative conceptual framework to tackle skill shortages in offsite construction industry: A scientometric approach. *Engineering, Construction and Architectural Management*, 29(10), 3917–3947. https://doi.org/10.1108/ECAM-04-2021-0287.

Saad, S., Alaloul, W.S., Ammad, S., Qureshi, A.H., Altaf, M. and Rasheed, K. (2020). Design Phase Carbon Emission Prediction Using a Visual Programming Technique. *2020 2nd International Sustainability and Resilience Conference: Technology and Innovation in Building Designs.* https://doi.org/10.1109/IEEECONF51154.2020.9319978.

Sarker, I.H. (2021). Deep learning: A comprehensive overview on techniques, taxonomy, applications and research directions. *SN Computer Science*, 2(6), 1–20. https://doi.org/10.1007/S42979-021-00815-1/FIGURES/6.

Sha, W., Guo, Y., Yuan, Q., Tang, S., Zhang, X., Lu, S., Guo, X., Cao, Y.-C. and Cheng, S. (2020). Artificial Intelligence to power the future of materials science and engineering. *Advanced Intelligent Systems*, 2(4), 1900143. https://doi.org/10.1002/AISY.201900143.

Smith, A. (2018). Franken-algorithms: The deadly consequences of unpredictable code. *The Guardian*, 30.

Unler, A. and Murat, A. (2010). A discrete particle swarm optimization method for feature selection in binary classification problems. *European Journal of Operational Research*, 206(3), 528–539. https://doi.org/https://doi.org/10.1016/j.ejor.2010.02.032.

Xia, Y. and Xu, Y. (2021). A transferrable data-driven method for IGBT open-circuit fault diagnosis in three-phase inverters. *IEEE Transactions on Power Electronics*, 36(12), 13478–13488. https://doi.org/10.1109/TPEL.2021.3088889.

Yamada, H., Liu, C., Wu, S., Koyama, Y., Ju, S., Shiomi, J., Morikawa, J. and Yoshida, R. (2019). Predicting materials properties with little data using shotgun transfer learning. *ACS Central Science*, 5(10), 1717–1730. https://doi.org/10.1021/acscentsci.9b00804.

Zhang, T. and Mo, H. (2021). Reinforcement learning for robot research: A comprehensive review and open issues. *International Journal of Advanced Robotic Systems*, 18(3), 17298814211007304. https://doi.org/10.1177/17298814211007305.

Zhang, X., Azhar, S., Nadeem, A. and Khalfan, M. (2018). Using building information modelling to achieve lean principles by improving efficiency of work teams. *International Journal of Construction Management*, 18(4), 293–300. https://doi.org /10.1080/15623599.2017.1382083.

CHAPTER

3

Artificial Intelligence in Construction Materials

Muhammad Imran Khan, Nasir Khan, Syed Saad and Muhammad Waqas Khan

Introduction

A vast area of computer science known as artificial intelligence (AI) develops intelligent robots by modeling human behavior. These devices make judgments that traditionally need human judgment and assist people in foreseeing and solving difficulties. Nowadays, AI technology is used in a wide range of sectors to create new tools and materials (Behnood & Golafshani, 2022). The intelligence system, a branch of computer science that develops and researches the best computation methods for addressing issues, serves the aim of employing AI techniques. A number of approaches can be adopted to develop the computational systems (Figure 3.1).

As an illustration, (Mashhadban et al., 2016) attempted to use the particle swarm optimization, (PSO), with artificial neural networks (ANNs) for prediction and modeling the mechanical characteristics of self-compacting concrete (SCC) reinforced with fiber. Okamura originally introduced SCC in 1997, after working issues with regular concrete, such as segregation, permeability, and bleeding, are avoided by using SCC and its compression under its weight. The authors of the study used AI to help them accomplish their objectives. The inaccuracy of the experimental methodologies used to assess the mechanical strength of concrete was shown in Ben Chaabene et al. (2020). In situations like cyclic loading, extreme thermo-mechanical loading, or any other sorts of loading, the concrete structures depict various behaviors. High temperatures and heavy loads cause the materials' nonlinear mechanical behavior to arise. Concrete's behavior is controlled by complex, highly temperature-dependent characteristics. In this situation, when the parameters are rarely stated and the concrete structure is subjected to significant loading, the use of AI is paramount.

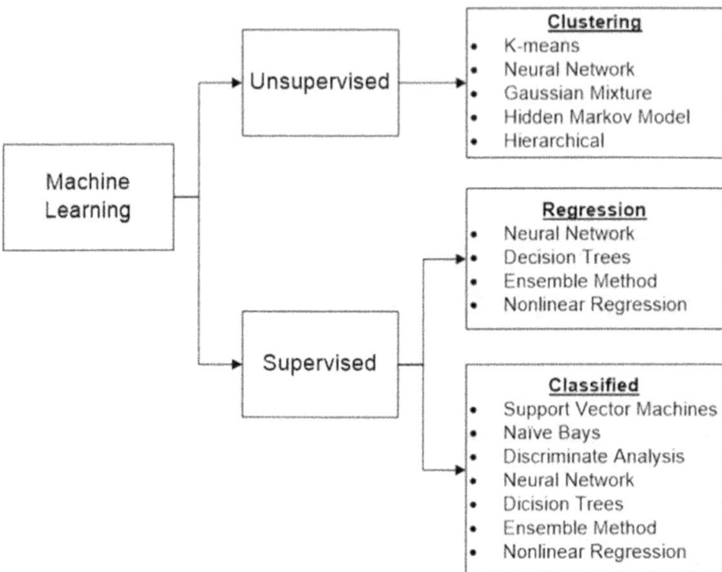

Figure 3.1: Most commonly adopted Machine Learning approaches.

A lot of research in the larger body of literature have looked more closely at this subject. Chou et al.'s (2016) goal was to improve the predictive accuracy of the compressive strength in high-performance concrete (HPC) by comparing several data-mining techniques. In this study, the viability of determining concrete's compressive strength utilizing additional ingredients as opposed to the water-to-cement ratio was examined. It was challenging to predict a specific compressive strength of concrete based on nonlinear relationships between the constituents, so it was wise to use quantitative analyses with five different data-mining techniques, including support vector machines (SVMs), ANN multiple regression, bagging regression trees (BRT), and multiple additive regression trees (MART) (Wang et al., 2022; Xiao et al., 2021). Bingöl in his article modeled the ANN approach used to determine the compressive strength of light-weight and semi-lightweight concretes that included pumice aggregate at greater temperatures (Bingöl et al., 2013). Pumice aggregate ratio, temperature and heating time were considered as input variables while conversely, concrete's compressive strength was used as an output parameter. Results obtained as the result of empirical testing and predictions of ANN model were compared where the values of the empirical method authenticated the values obtained as a result of the ANN model.

Smart Materials and Their Properties

These materials may be natural or man-made which shows their properties as desired in response to the environmental changes. Based on the research, these materials can change their properties when they are exposed to external stimuli in a predefined manner. The stimuli can include electric current, electromagnetic waves, pressure, friction, ultraviolet radiation, etc., while in response to these, properties like size, shape, color and ductility, etc. may change. The superiority of smart materials over the conventional materials and rapid return on investment have led the use of smart materials to a next level across the last ten years in the building business (Alaloul et al., 2020). Some of the reasons behind this boost are given below.

i. It can provide a greater precision and rapid response in sensors applications.
ii. It can provide crucial information regarding service life and durability, i.e., deterioration and strength.
iii. It offers more resistance against fir, fracture, fatigue and chemicals.
iv. It helps in implementing energy efficient and environmentally friendly building designs.

Some of smart materials with their advantages, disadvantages, and their applications have been shown in Table 3.1.

Piezoelectric Materials

Materials that are capacitive alter when a force is applied. Electric pulses appear when any impact is given to these materials, and vice versa. A direct and converse effect are terms used to describe the aforementioned events. A change in the material's size results in the production of an electric current. From civil engineering to aeronautics, these materials are widely used. Applications include the following:

a. Microscale energy harvesting, which makes use of piezoelectric transformers;
b. Usage in automotive as well as sensors and actuators as accelerometers;
c. Clocks use consistency as an indicator;
d. Ultrasound equipment and underwater detectors also use frequency;
e. A camera lens with a piezo motor is employed in a medical application.

Transparent Concrete

Cement is one of the more commonly employed building materials. It can be made more environmentally friendly and useful in a variety of ways. By including translucent materials into the mix or incorporating fiber into the concrete, transparent or translucent concrete that efficiently transmits light may be created. Additionally, it is the newest and greenest concrete to date. Transparent concrete constructions offer a quality polish and aesthetically pleasing structures. It is crucial to design structures that are economical, environmentally beneficial, and structurally sound to produce novel materials. Transparent concrete is both more affordable and environmentally friendly. During the day, it may create light by absorbing light, and at night, it can glow with a glimmer. Translucent

Table 3.1: Smart materials and their applications

Name	Advantages	Disadvantages	Applications
Piezoelectric materials	High sensitivity and stability	Not applicable to calculations that are static.	Infrastructure and roads
Materials with magneto restriction	Strong energy density and natural hardness.	The complexity of the substances increases.	Damper, sensors, energy storage (Olabi & Grunwald, 2008).
Transparent concrete	Environment friendly, transmit light effectively.	Casting is difficult, High initial cost.	Any construction (Kashiyani et al., 2013).
Self-healing coating	Self-heals, repairs, and reconstructs on its own.	Time consuming, dense resin should be avoided.	Cementitious materials, corrosion protection (Koh et al., 2014).
Retain their form alloys	High strength, flexible to begin with.	Pricey to produce.	Tremor resistant structures (Mirzaeifar et al., 2012).
Aerogels	Used as insulation because of its porous nature, efficient and adaptable.	Reduced clarity and surface area.	As an insulation in buildings (Riffat & Qiu, 2013).
Self-sensing concrete	Perceive a slight variation in stress and strain.	Cannot perceive the variations at early age.	Building structure (Han et al., 2015).
Smart bricks	Sense sound levels, stress, moisture, vibrations and chemical.	High initial cost and maintenance.	Building structure (D'Alessandro et al., 2018).
Smart wrap	Temperature control and high strength.	Not yet fully developed.	Walls and partitions wall material replacement.
Smart Glass	Cost-effective, amends the intensity of heat and light.	High initial cost.	Automobiles, Buildings (Ke et al., 2019).
A shift in pitch	Latent heat capture and release; conserving energy.	Fiscal and mutually beneficial must be determined.	Added into concrete, thermal storage (M.I. Khan et al., 2019).

aluminum structures appear brighter during the day. Due to the imbedded carbon fibers, they have considerable benefit in economical construction and they may reduce lighting costs and offer insulation during the winter (Saad, Salah Alaloul, et al., 2020). They may also be utilized in non-structural members, interior components, paving, and walls.

Self-healing Concrete

The substance which is most frequently used in building nowadays because of its flexibility and simplicity of manufacture is cement. However, it has a significant flaw in that it is prone to cracking notwithstanding any bending strain, rapid heat change, or incorrect setting, which lowers its strength and endurance. Because "Bacillus pseudofirmus" bacteria are present, which eat calcium lactate to produce calcium carbonate, they are also known as microbial concrete. Cracks and deformations are repaired by the limestone. During batching and manufacture, these microorganisms are included into the concrete. The material's tenacity and lasting durability are increased by this self-healing process. Concrete cannot absorb water or gases because of the coating of calcium carbonate.

Shape Memory Metals

The reason why these substances are frequently referred to as "transforming substances" are because they can 'memorize' their original shape and 'remember' how to return to that shape after being distorted. Metals having shape-memory properties can be utilized for bridges because they can withstand stress before returning to their original shape after being released. If a component constructed of a shape memory alloy is distorted by a natural disaster, such as a tornado or a strong wind gust, it will return to its former shape after some time has passed. Due to a lack of study and safety concerns, these materials are not yet widely employed in the building sector. Extensive study and time are needed to realize many of the potential uses. The structural features of these alloys are low weight and excellent yield and plastic strength. Smart reinforced concrete, earthquake-resistant buildings, pipe couplings, etc. might all benefit from the usage of shape-shifting materials. There is surprisingly little study history for these materials despite their impressive structural characteristics. These materials are costlier than structural steel and may corrode fast. Polluting the environment is an inevitable byproduct of extracting such materials.

Self-sensing Concrete

Electric resistance is a property of this type of concrete which makes this mix capable of sensing small changes in stress and strain. Small carbon fibers are incorporated in self-sensing concrete which is why it is sometime referred as carbon fiber reinforced concrete. The detection system that locates the electrodes on the framework of this type of ceramic is made of carbon fibers (Saad, Alaloul, et al., 2020). Due this property of the mix, electric probe can be mounted anywhere on the surface to measure the electric resistivity. Steel fibers can be substituted for carbon fiber filaments in certain smart concrete carbon to improve its properties. This insertion can be very useful to sense the changes in stress in the sub-structure before any earthquake.

Smart Bricks

These bricks may be used to keep a record of numerous structural parameters extremely effectively. To detect strain, electrodes and conductive nano filler are implanted in certain bricks (Downey et al., 2018), while others are enrooted that comprise three key technological parts: a transmitter and receiver, data controllers, and detectors (Engel et al., 2023). When utilized in a construction, these bricks can detect a variety of forces and factors, including changes in sound levels, stress, moisture content, chemical changes, vibration, type of force, and so on (D'Alessandro, Ubertini, et al., 2018). Sensors of various sorts are used for various reasons. These capabilities will be useful in unanticipated scenarios such as fire threats, earthquake damage assessment, and so on (Meoni et al., 2019). By providing vital information on the environmental conditions of the structure, the use of these bricks might enhance tenant health as well as safety (D'Alessandro, Meoni, et al., 2018).

Smart Glass

Similar to Smart Wrap, Smart glasses are economical and can change their properties depending on the situation. They have the capability of changing the quantity of heat, glare and light that passes through them. They can be installed in windows, doors, sun blinds and sunroofs, etc. Smart glasses are classified into two types: active and passive. Passive smart glasses modify their features on their own when activated by sunshine and heat. Photochromic and thermochromics glasses are common examples of such glasses. Active glasses convert from transparent to opaque with a click of a button offering privacy as desired. However, passive glasses are more environment friendly than active glasses, but on the other hand, they offer more detailed settings than passive glasses in terms of heat and light transmission. Two glass or clear plastic surfaces with unique conductivity coatings on the panel inside make up panels based on Surge Protective Devices (SPD) technology. The coating film is made up of minute particles that are 'suspended' between two surfaces covered in glass according to a chemical design.

AI Assisted Material Design and Development

A significant motivation behind employing machine learning (ML) methods pertaining to the study of materials is to expedite the identification of innovative materials. As materials science research progresses, there is a continuous accumulation of data from both experimental and simulation studies, establishing a robust data infrastructure that facilitates the widespread utilization use of AI techniques in the realm of materials research and development. ML has revolutionized simulation technology and materials prediction (Saad, Alaloul, et al., 2020), bringing up potential breakthroughs from the field of civil construction to the development of Lithium-ion batteries. Takahashi & Tanaka (2016) addressed the ML approach in the anticipation of qualities of material. Big data applications and ML approaches can be adopted to train machines based on big datasets of materials to predict new materials of desired qualities. A study by A. Khan et al. (2009) adopted ML to access the nonlinear relationship between mechanical characteristics, temperature, and composition of nanocomposites. Other than civil engineering, ML has proven its capabilities in other fields like solar energy and photovoltaic sector; medical diagnosis have taken a special interest in the new frontier of materials prediction. The benefits of data-driven simulation and algorithmic learning in mathematical mechanics and composites design has become more widely acknowledged. Hamdia et al. (2015) started using ML to determine the best way to construct sandwich structures and forecast their fracture toughness. They believe, using ML and large-scale data sets may usher in a new paradigm for developing materials for a specific use. Beside this, Kirchdoerfer & Ortiz (2016) present two mathematical modeling case studies that illustrate the application of ML and big data applications as a powerful computing tool.

Research attempted to foresee the mechanical features of cement-based materials while developing ML models. Nazari & Sanjayan (2015) employed an SVM to estimate the geopolymer's compressive strength. To determine the SVM's parameters, four well-known PSOs are examples of optimization approaches that were used: genetic algorithms, optimization of ant colonies algorithms, imperialist competitive algorithms, and artificial colonies of bees' algorithmic optimization. Based on the errors and coefficient of determination calculated for every case, genetic algorithm and imperialist competitive approach performed the best while predicting the geopolymer's compressive strength using the SVM.

Ramadan and Nehdi (2017) employed a hybrid genetic algorithm-artificial neural network (ANN) technique to estimate the intrinsic self-healing capability of concrete. Within this framework, a genetic algorithm was created to serve as a probabilistic optimization tool for determining the initial optimum weights and biases in the neural network. The W/C ratio, cement %, type of SCM, amount of SCM, crystalline additive, expansive additive, and bio-healing material were employed as input factors, with fracture width as the output parameter for the concrete's ability to repair itself. It was found after assessing the performance validation criteria that the model could reflect the performance impacts of several self-healing agents of the concrete.

The integration of AI techniques with Finite Element Method (FEM) and Molecular Dynamics (MD) is a common practice in the field of nanomaterial characterization. In their study on Al-matrix nanocomposites, Shabani & Mazahery (2012) aimed to evaluate the impact on the mechanical characteristics of the volume % of alumina nanoparticles of an aluminum-silicon (Al-Si) matrix. The observed reduction in elongation and concurrent rise in tensile and yield strength resulting from the volume percentage of Al_2O_3 that rose suggests that this augmentation has the potential to enhance strength characteristics while diminishing flexibility. Vijayaraghavan and colleagues (2014) introduced a novel simulation method that combines molecular dynamics (MD) with AI to examine the heat transport properties of carbon nanotubes. This novel method marks the first investigation into the effects of chirality, size, and voids influence carbon tubes' thermal resistance. This information was used as input for an AI cluster that used genetic programming as its main methodology to build a direct correlation between the thermal transport of carbon nanotubes (CNTs) and variables including system size, chirality, and vacancy defect concentration (Vijayaraghavan et al., 2014).

More focus is being placed on the mechanical characteristics of metals and their composites, and among them, static strength is the mechanical property that is most frequently anticipated. A back-propagation neural network (BPNN) was used by Altinkok & Koker (2004) to foresee the tensile strength of aluminum reinforced, using Al_2O_3 and silicon dioxide granules. The bending strength was estimated using experiments to create a training and testing dataset based on the particle size reinforced phase. The model showed that the bending strength increased with a reduction in the size of reinforced SiC particles based on the data's hidden patterns. By controlling the size of the reinforced SiC particle, this innovation aids the industry in creating a novel material with the necessary bending strength.

Later, a similar study using a variety of training techniques was carried out by Koker et al. (2007). Levenberg-Marquardt outperformed the other approaches in this study for predicting the bending characteristics of the aluminum metal matrix. Robust back-propagation, quasi-newton, and variable learning back-propagation were also used. While examining the impact of the stair welding process Jayaraman et al. (2008) came to the conclusion that rotational speed as a process factor among axial force and welding speed influence the strength. On the basis of the dataset produced from the finite element simulation, in another study G. Chen et al. (2019) built regression and classification models utilizing ANN to map the input variables and forecast the strength of the particle reinforced metal matrix.

To calculate the tribological characteristics of a magnesium metal matrix composite augmented with reduced graphene oxide, Kavimani & Prakash (2017) used a Taguchi coupled ANN technique. The composite's tribological performance was represented by the particular wear rate, which was used as the output variable. The applied load, the four input variables employed were sliding velocity, sliding distance, and reinforcement weight to reinforcement weight ratio. The findings indicated that the specific composite wear rate was more significantly impacted by load. Genel et al. (2003) developed a multilayer feed-forward ANN to simulate the tribological behavior of zinc-aluminum composites reinforced with short alumina fibers. Predicting the creep and fatigue behaviors of metallic materials is a remarkable application of AI technology.

To anticipate the creepy actions of a spinning composite disc that comprised of aluminum reinforced with silicon carbide particles (Al-SiCp) at high temperatures, a model using ANNs was created. They calculated the creep response for various mixtures of temperature, particle concentration, and particle size. When compared to identical analytical values, the findings produced by the ANN tended to be slightly more cautious, highlighting the necessity of including a safety margin in such predictions (Gupta et al., 2007). An ANN model was developed in a different study to predict how well CK45 mild steel will perform during fatigue. The ANN was utilized to calculate fatigue life using input layers made up of stress amplitude and coating thickness. The Finite Element Method (FEM) results were utilized to compare the projected performance, and fatigue life tests were employed to collect the training data for the ANN model. The outcomes demonstrated that the ANN model performed better than the FEM model. In their 2017 study, Maleki and Kashyzadeh used a property-based model with inputs for ultimate tensile strength and elongation to failure. However, their process-based model took into account a variety of processing and post-processing variables (Maleki & Kashyzadeh, 2017).

In recent decades, the performances of metallic materials other than their mechanical properties have also been explored utilizing AI approaches. Sivasankaran et al. (2009) employed an ANN model to analyze the workability characteristics of Al-SiC metal matrix composites produced using powder metallurgy (Sivasankaran et al., 2009). Using the Levenberg-Marquardt learning method, the feed-forward backpropagation ANN model was trained. A wide range of traditional statistical criteria were used to evaluate the effectiveness of the ANN model.

Numerous studies have examined various applications of AI approaches to the modeling of polymers and their composites. The most significant and frequently discussed component of these applications is the prediction of polymer characteristics, whereas the majority of research has focused on mechanical aspects. A common prediction technique includes a training phase that includes model testing, modifying parameters and weights, and a predictive phase that uses the optimized and trained model to forecast the outcome of the prediction and foretell future outcomes.

In their study, Pidaparti & Palakal (1993) used a neural network (BPNN) model, introduced a back-propagation for predicting their nonlinear stress-strain properties of graphite-epoxy laminates that they anticipated. Using a dataset of 959 data points, this model was created. Two intermediary layers and one output layer made up the network used in this investigation. The output layer comprised a single node representing total strain, whereas the intermediate layers contained three nodes representing fiber-angles, beginning stress, and incremental stress. In the light of the aforementioned circumstances, it is imperative to consider the potential implications and ramifications in the study conducted by Jiang et al.

(2008) utilized an ANN model to make predictions regarding the mechanical and wear characteristics of polyamide composites reinforced with short fibers. This study utilized two datasets for the purpose of training the network. The input elements encompassed material compositions, testing settings, and manufacturing methods. The mechanical quality and wear characteristics were included as output parameters. There has been considerable interest in the investigation of the dielectric constant and glass transition temperature in many studies.

AI in Material Synthesis and Characterization

Material synthesis, a pivotal aspect of civil engineering, involves designing new materials with specific properties to enhance structural performance, durability, and sustainability. ML and AI techniques have revolutionized this process by expediting the discovery and development of advanced materials. Material synthesis explains the physical and chemical properties of civil engineering material providing a platform for developing an optimum mix with strength properties as per requirement. In this regard, researchers have adopted the use of computers approaches to forecast various characteristics of various civil engineering substances (Abdul Hannan Qureshi et al., 2020). Cao (2023) adopted ensemble trees for predicting the porosity of concrete mix based on 240 observations collected from literature. Seventy-five percent of the data points among these were selected at random to train the ensemble. The rest of the 25% are models used to validate the model. Based on the performance of the established models it was discovered that ML is very effective at forecasting the porosity of concrete. Huang et al. (2020) carried out research on the application of random forest (RF) and beetle antennae search (BAS) to access the permeability of concrete using 36 samples prepared in the lab with four key input parameters. BAS and RF hybrid model can be adopted to design a concrete with desired permeability based on the models created for this investigation. Performance considering the designs created in this study was validated by using R and $E_{RMS;}$ the sensitivity analysis shows that the permeability of concrete is affected the most by the cement to aggregate ratio.

Another study by Kassa & Wubineh (2023) presents the use of machine RF. ANN, decision tree (DT), and linear regression (LR) were used to predict the soil's California Bearing Ratio (CBR), which was calculated using seven input variables. The authors mapped these variables to predict the CBR by splitting the data in 80:20 for modeling instruction and evaluation. The performers result models created was validated by applying R^2, RMSE, MAE, and MSE. Results show that the RF algorithms outperform the rest of the algorithms while predicting the CBR of the soil. Nguyen et al. (2020) adopted Deep Residual Networks and Deep Neural Networks to access the constructive strength of geopolymer cement mixed with the fly ash. From the results, it was observed that the Deep Residual Networks have a better predicting power compare to the Deep Neural Network.

The methodology adopted in the study for hybrid model developed can be seen in Figure 3.2. The same research was done by Dao et al. (2019) with no fly ash using ANNs with Adaptive Neuro-Fuzzy Inference (ANFIS) where ANFIS outclassed the ANN by providing higher accuracy as compared to ANN. Another study by Azimi-Pour et al. (2020) used SVM algorithms with both linear and nonlinear parameters a number of kernal settings to forecast high volume-based strength predictions of self-compacting concrete. Based on the performance, it was observed that the RBF-based SVM system kernal performs the best to predict the compressive strength of self-compacting asphalt that has been altered with a lot of fly ash.

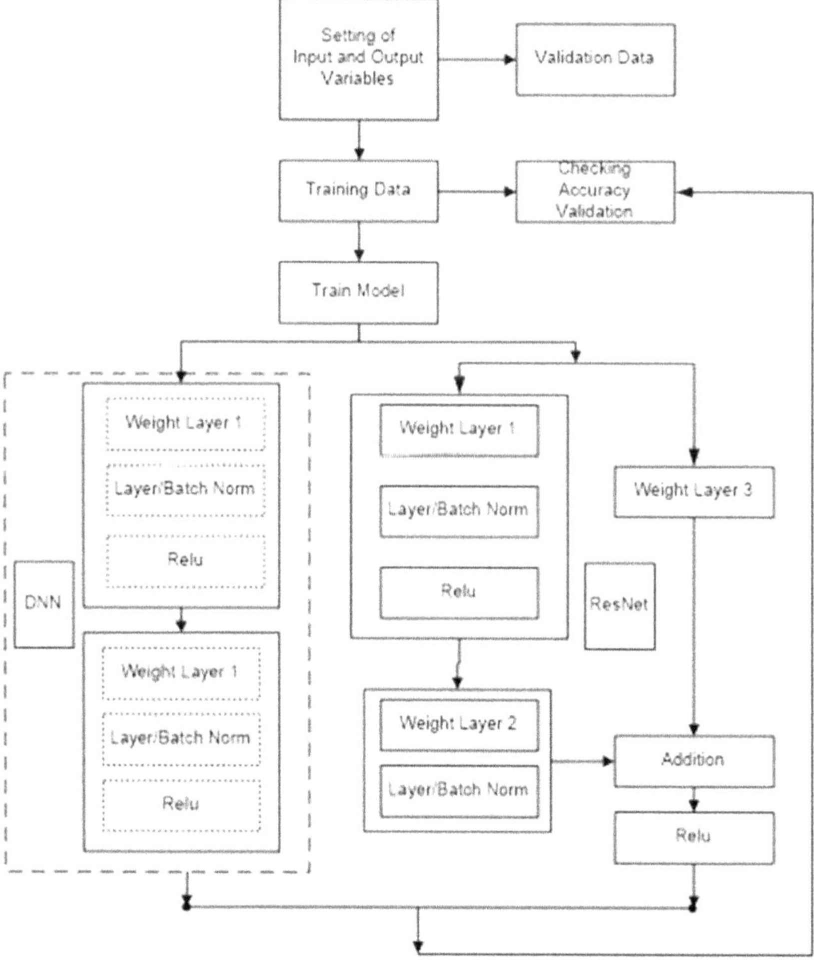

Figure 3.2: Flow chart for DNN and ResNet for predicting compressive strength.

AI in Construction Material Testing and Analysis

The construction industry faces numerous challenges that hinder its growth and result in significantly lower productivity levels compared to industries like manufacturing. In fact, it is one of the least digitized sectors globally, with a deep-rooted resistance to change acknowledged by most stakeholders. The industry's lack of digitization and heavy reliance on manual processes make project management more complex and unnecessarily cumbersome. Moreover, the limited adoption of digital expertise and technology within construction has been associated with cost inefficiencies, project delays, subpar quality, uninformed decision-making, and overall poor performance in productivity (Abbasi et al., 2022), health, and safety. Recently, it has become evident that the construction industry must embrace digitization and swiftly enhance its technological capabilities. AI methods have been utilized in a variety of construction industries as illustrated in the Figure 3.3.

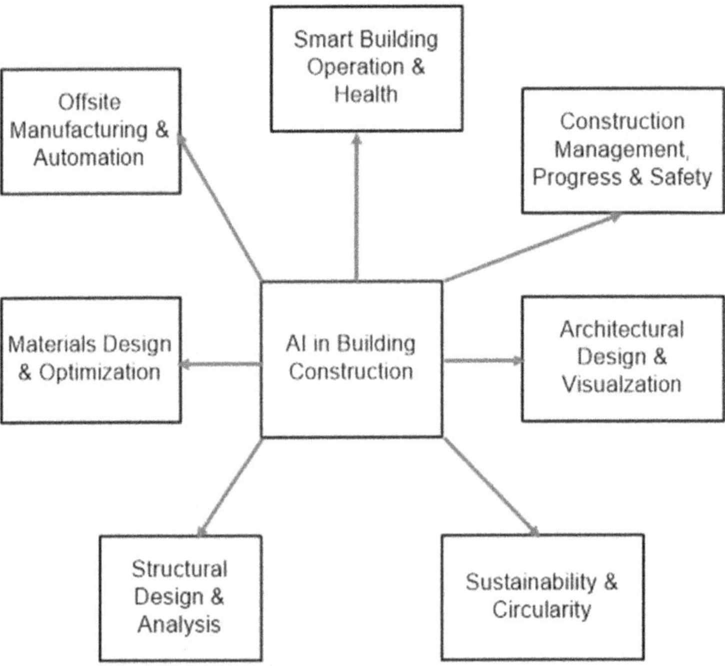

Figure 3.3: Application of AI in the construction industry.

Since the last few years, applications for computer learning as well as computational intelligence (also called AI) in construction material testing and analysis are gaining momentum in research and industry (Saad et al., 2022). AI in construction material testing and analysis has emerged as a transformative technology, enhancing efficiency and accuracy in the construction industry. By leveraging ML algorithms and computer vision, AI enables automated quality control, detecting defects, and ensuring high-quality materials are used. Predictive analysis based on historical data empowers construction professionals to optimize material usage and plan maintenance, while AI-driven material property prediction facilitates precise material selection and performance evaluation. This integration of AI into construction material testing and analysis streamlines processes, improves project outcomes, and contributes to safer and more sustainable construction practices.

AI in Building/Concrete Materials Testing and Analysis

Building time, endurance, power, energy effectiveness, emissions, appearance, and the warmth of a structure are all influenced by the choice of construction materials. By employing AI techniques, it is possible to create and develop high-performance materials and composites that take into account these factors. Researchers have conducted investigations to develop models that incorporate ML in order to predict mechanical properties. These efforts aim to reduce material consumption, cost, and testing time, thereby streamlining the overall process.

Numerous research studies have been conducted to forecast concrete behavior by utilizing various AI techniques, for example, ANNs, DL, SVMs, GA (Genetic Algorithms), and fuzzy logic are other applications. The majority of these applications focus on using ANN or SVM to anticipate practical scenarios, such as properties, for example, tensile force, compressive durability, and elastic modulus. Some studies extend their focus to concrete exposed to extreme environments, such as high temperatures. Additionally, a few investigations have employed ANN to predict concrete strength when incorporating nanomaterials, fly ash, and other mineral admixtures, as well as powdered slag from the blast (Bui et al., 2018).

Bui et al. (2018) utilized a mix of ANN and modified firefly algorithm (MFA) (known as MFA-ANN method) for forecasting superior-performing concrete's compressive and tensile strength. These two properties are highly dependent and present nonlinear behavior with the constituents of concrete. The proposed method showed lower RMSE, MAE, and MAPE compared with other methods and indicated its improved performance. Moreover, Higher R value of proposed model MFA-ANN is higher than other methods which indicates its higher significance.

With technological advancements, making use of ML techniques prices have risen due to widespread use in addressing a diverse range of engineering challenges. Various optimization techniques have been established in different technical and multidisciplinary fields to achieve outstanding levels of performance. For example, Iqbal et al. (2021) developed using the forest randomization model to explore how input parameters (Bar size and percentage of volume) affect TSR of FRP. Model comparison using RSM and ANN were also performed to investigate which tool is more significant compared to the others. The results revealed that ANN model gives a higher value of R compared with RSM model indicating its higher significance and higher accuracy as shown in Table 3.2. The outcomes demonstrated the dependable foreseen of TSR of GFRP rebar utilizing the RFs model in an acidic setting. However, the traditional method of evaluating the impact of the bar size and volume proportion involved standard parameter testing and sensitivity analysis. This approach entails assessing data that is generated using the learned model, where one variable is varied while others are maintained at average values. A limitation of this conventional sensitivity analysis is its dependence on simulated data rather than capturing the interaction patterns found in the complete dataset. Similarly, in another study, the gene expression programming (GEP) model was employed to investigate the long-term remaining tensile strength.

Table 3.2: Model terms of RSM and ML analysis

Origin	Financial Index	Cost
ANOVA RSM	Template	Major
	Squared-sum model	16897.78
	F-value	18.11
	p-value	<0.0005
	Improper fit	not important (p = 0.9985)
	R	0.854
	Modified R2	0.70
	Predicted R2	0.66
	Equation accuracy	19.76
Light GBM (Data for training)	R	0.93
	RMSE	4.854
	MAE	3.073
Light GBM (Test data)	R	0.868
	RMSE	6.199
	MAE	4.078

Currently, there is extensive research focused on utilizing computer learning techniques to foresee concrete robustness. These techniques encompass ANNs, regression models, trees of choices, and assistance vector machines, and various individuals. Implementing these methods can offer notable benefits by enabling faster and more cost-effective estimations of concrete compressive strength. Chen et al. (2018) employed both ANN and SVM algorithms to forecast concrete's compressible strength (Chen et al., 2018). According to this study, the anticipated sturdiness strengths derived from the newly created SVM pattern exhibited a strong alignment with the empirical results. This alignment suggests that the

SVM model holds promise for forecasting the concrete-based materials' strength in compression utilized in challenging conditions. When contrasted with models based on ANN, the SVM algorithms displays superior performance.

References

Abbasi, M.F., Bilal, M. and Rasheed, K. (2022). Role of Human Intuition in AI Aided Managerial Decision Making: A Review. *2022 International Conference on Decision Aid Sciences and Applications, DASA 2022*. https://doi.org/10.1109/DASA54658.2022.9765153.

Abdul Hannan Qureshi, Wesam Salah Alaloul, Manzoor, B., Musarat, M.A., Saad, S. and Ammad, S. (2020). Implications of Machine Learning Integrated Technologies for Construction Progress Detection Under Industry 4.0 (IR 4.0). *2020 Second International Sustainability and Resilience Conference: Technology and Innovation in Building Designs, 0*. https://doi.org/10.1109/IEEECONF51154.2020.9319974.

Alaloul, W.S., Saad, S. and Qureshi, A.H. (2020). Construction sector: IR 4.0 applications. *In:* C.M. Hussain and P. Di Sia (Eds.), *Handbook of Smart Materials, Technologies, and Devices: Applications of Industry 4.0*, pp. 1–50. Springer International Publishing. https://doi.org/10.1007/978-3-030-58675-1_36-1.

Altinkok, N. and Koker, R. (2004). Neural network approach to prediction of bending strength and hardening behaviour of particulate reinforced (Al–Si–Mg)-aluminium matrix composites. *Materials & Design*, 25(7), 595–602. https://doi.org/10.1016/j.matdes.2004.02.014.

Azimi-Pour, M., Eskandari-Naddaf, H. and Pakzad, A. (2020). Linear and nonlinear SVM prediction for fresh properties and compressive strength of high volume fly ash self-compacting concrete. *Construction and Building Materials*, 230, 117021. https://doi.org/10.1016/j.conbuildmat.2019.117021.

Behnood, A. and Golafshani, E.M. (2022). Artificial Intelligence to model the performance of concrete mixtures and elements: A review. *Archives of Computational Methods in Engineering*, 29(4), 1941–1964. https://doi.org/10.1007/s11831-021-09644-0.

Ben Chaabene, W., Flah, M. and Nehdi, M.L. (2020). Machine learning prediction of mechanical properties of concrete: Critical review. *Construction and Building Materials*, 260, 119889. https://doi.org/10.1016/j.conbuildmat.2020.119889.

Bingöl, A.F., Tortum, A. and Gül, R. (2013). Neural networks analysis of compressive strength of lightweight concrete after high temperatures. *Materials & Design (1980–2015)*, 52, 258–264. https://doi.org/10.1016/j.matdes.2013.05.022.

Bui, D.-K., Nguyen, T., Chou, J.-S., Nguyen-Xuan, H. and Ngo, T.D. (2018). A modified firefly algorithm-artificial neural network expert system for predicting compressive and tensile strength of high-performance concrete. *Construction and Building Materials*, 180, 320–333. https://doi.org/10.1016/j.conbuildmat.2018.05.201.

Cao, C. (2023). Prediction of concrete porosity using machine learning. *Results in Engineering*, 17, 100794. https://doi.org/10.1016/j.rineng.2022.100794.

Chen, G., Wang, H., Bezold, A., Broeckmann, C., Weichert, D. and Zhang, L. (2019). Strengths prediction of particulate reinforced metal matrix composites (PRMMCs) using direct method and artificial neural network. *Composite Structures*, 223, 110951. https://doi.org/10.1016/j.compstruct.2019.110951.

Chen, H., Qian, C., Liang, C. and Kang, W. (2018). An approach for predicting the compressive strength of cement-based materials exposed to sulfate attack. *PLoS One*, 13(1), e0191370.

Chou, P.-Y., Tsai, J.-T. and Chou, J.-H. (2016). Modeling and optimizing tensile strength and yield point on a steel bar using an artificial neural network with Taguchi particle swarm optimizer. *IEEE Access*, 4, 585–593. https://doi.org/10.1109/ACCESS.2016.2521162.

D'Alessandro, A., Ubertini, F., Downey, A., Laflamme, S. and Meoni, A. (2018). Strain monitoring in masonry structures using smart bricks. *In:* H. Sohn (Ed.), *Sensors and Smart Structures Technologies for Civil, Mechanical, and Aerospace Systems 2018*, p. 66. SPIE. https://doi.org/10.1117/12.2297526.

Dao, D., Ly, H.-B., Trinh, S., Le, T.-T. and Pham, B. (2019). Artificial Intelligence approaches for prediction of compressive strength of geopolymer concrete. *Materials*, 12(6), 983. https://doi.org/10.3390/ma12060983.

Downey, J.E., Schwed, N., Chase, S.M., Schwartz, A.B. and Collinger, J.L. (2018). Intracortical recording stability in human brain–computer interface users. *Journal of Neural Engineering*, 15(4), 046016. https://doi.org/10.1088/1741-2552/aab7a0.

Engel, Y., Mannor, S. and Meir, R. (2004). The kernel recursive least-squares algorithm. *IEEE Transactions on Signal Processing*, 52(8), 2275–2285. https://doi.org/10.1109/TSP.2004.830985.

Genel, K., Kurnaz, S.C. and Durman, M. (2003). Modeling of tribological properties of alumina fiber reinforced zinc–aluminum composites using artificial neural network. *Materials Science and Engineering: A*, 363(1–2), 203–210. https://doi.org/10.1016/S0921-5093(03)00623-3.

Gupta, V.K., Kwatra, N. and Ray, S. (2007). Artificial neural network modeling of creep behavior in a rotating composite disc. *Engineering Computations*, 24(2), 151–164. https://doi.org/10.1108/02644400710729545.

Hamdia, K.M., Lahmer, T., Nguyen-Thoi, T. and Rabczuk, T. (2015). Predicting the fracture toughness of PNCs: A stochastic approach based on ANN and ANFIS. *Computational Materials Science*, 102, 304–313. https://doi.org/10.1016/j.commatsci.2015.02.045.

Han, B., Ding, S. and Yu, X. (2015). Intrinsic self-sensing concrete and structures: A review. *Measurement*, 59, 110–128. https://doi.org/10.1016/j.measurement.2014.09.048.

Huang, J., Duan, T., Zhang, Y., Liu, J., Zhang, J. and Lei, Y. (2020). Predicting the permeability of pervious concrete based on the beetle antennae search algorithm and random forest model. *Advances in Civil Engineering*, 2020, 1–11. https://doi.org/10.1155/2020/8863181.

Iqbal, W., Tang, Y.M., Chau, K.Y., Irfan, M. and Mohsin, M. (2021). Nexus between air pollution and NCOV-2019 in China: Application of negative binomial regression analysis. *Process Safety and Environmental Protection*, 150, 557–565. https://doi.org/https://doi.org/10.1016/j.psep.2021.04.039.

Jayaraman, M., Sivasubramanian, R., Balasubramanian, V. and Lakshminarayanan, A.K. (2008). Prediction of tensile strength of friction stir welded A356 cast aluminium alloy using response surface methodology and artificial neural network. *Journal for Manufacturing Science and Production*, 9(1–2), 45–60. https://doi.org/10.1515/IJMSP.2008.9.1-2.45.

Ji, G., Li, F., Li, Q., Li, H. and Li, Z. (2010). Prediction of the hot deformation behavior for Aermet100 steel using an artificial neural network. *Computational Materials Science*, 48(3), 626–632. https://doi.org/10.1016/j.commatsci.2010.02.031.

Jiang, Z., Gyurova, L., Zhang, Z., Friedrich, K. and Schlarb, A.K. (2008). Neural network based prediction on mechanical and wear properties of short fibers reinforced polyamide composites. *Materials & Design*, 29(3), 628–637. https://doi.org/10.1016/j.matdes.2007.02.008.

Kashiyani, B.K., Raina, V., Pitroda, J. and Shah, B.K. (2013). A study on transparent concrete: A novel architectural material to explore construction sector. *International Journal of Engineering and Innovative Technology (IJEIT)*, 2(8), 83–87.

Kassa, S.M. and Wubineh, B.Z. (2023). Use of machine learning to predict California bearing ratio of soils. *Advances in Civil Engineering*, 2023, 1–11. https://doi.org/10.1155/2023/8198648.

Kavimani, V. and Prakash, K.S. (2017). Tribological behaviour predictions of r-GO reinforced Mg composite using ANN coupled Taguchi approach. *Journal of Physics and Chemistry of Solids*, 110, 409–419. https://doi.org/10.1016/j.jpcs.2017.06.028.

Ke, Y., Chen, J., Lin, G., Wang, S., Zhou, Y., Yin, J., Lee, P.S. and Long, Y. (2019). Smart windows: Electro-, thermo-, mechano-, photochromics, and beyond. *Advanced Energy Materials*, 9(39), 1902066. https://doi.org/10.1002/aenm.201902066.

Khan, A., Shamsi, M.H. and Choi, T.-S. (2009). Correlating dynamical mechanical properties with temperature and clay composition of polymer-clay nanocomposites. *Computational Materials Science*, 45(2), 257–265. https://doi.org/10.1016/j.commatsci.2008.09.027.

Khan, M.I., Huat, H.Y., Dun, M.H. bin M., Sutanto, M.H., Jarghouyeh, E.N. and Zoorob, S.E. (2019). Effect of irradiated and non-irradiated waste PET based cementitious grouts on flexural strength of semi-flexible pavement. *Materials*, 12(24), 4133. https://doi.org/10.3390/ma12244133.

Kirchdoerfer, T. and Ortiz, M. (2016). Data-driven computational mechanics. *Computer Methods in Applied Mechanics and Engineering*, 304, 81–101. https://doi.org/10.1016/j.cma.2016.02.001.

Koh, E., Kim, N.-K., Shin, J. and Kim, Y.-W. (2014). Polyurethane microcapsules for self-healing paint coatings. *RSC Adv.*, 4(31), 16214–16223. https://doi.org/10.1039/C4RA00213J.

Koker, R., Altinkok, N. and Demir, A. (2007). Neural network based prediction of mechanical properties of particulate reinforced metal matrix composites using various training algorithms. *Materials & Design*, 28(2), 616–627. https://doi.org/10.1016/j.matdes.2005.07.021.

Maleki, E. and Kashyzadeh, K.R. (2017). Effects of the hardened nickel coating on the fatigue behavior of CK45 steel: Experimental, finite element method, and artificial neural network modeling. *Iranian Journal of Materials Science and Engineering*, 14(4), 81–99. https://doi.org/10.22068/ijmse.14.4.81.

Mashhadban, H., Kutanaei, S.S. and Sayarinejad, M.A. (2016). Prediction and modeling of mechanical properties in fiber reinforced self-compacting concrete using particle swarm optimization algorithm and artificial neural network. *Construction and Building Materials*, 119, 277–287. https://doi.org/10.1016/j.conbuildmat.2016.05.034.

Meoni, A., D'Alessandro, A., Cavalagli, N., Gioffré, M. and Ubertini, F. (2019). Shaking table tests on a masonry building monitored using smart bricks: Damage detection and localization. *Earthquake Engineering & Structural Dynamics*, 48(8), 910–928. https://doi.org/10.1002/eqe.3166.

Mirzaeifar, R., DesRoches, R., Yavari, A. and Gall, K. (2012). Coupled thermo-mechanical analysis of shape memory alloy circular bars in pure torsion. *International Journal of Non-Linear Mechanics*, 47(3), 118–128. https://doi.org/10.1016/j.ijnonlinmec.2012.01.007.

Nazari, A. and Sanjayan, J.G. (2015). Modeling of compressive strength of geopolymer paste, mortar and concrete by optimized support vector machine. *Ceramics*

International, 41(9), 12164–12177. https://doi.org/10.1016/j.ceramint.2015.06.037.

Nguyen, K.T., Nguyen, Q.D., Le, T.A., Shin, J. and Lee, K. (2020). Analyzing the compressive strength of green fly ash based geopolymer concrete using experiment and machine learning approaches. *Construction and Building Materials*, 247, 118581. https://doi.org/10.1016/j.conbuildmat.2020.118581.

Olabi, A.G. and Grunwald, A. (2008). Design and application of magnetostrictive materials. *Materials & Design*, 29(2), 469–483. https://doi.org/10.1016/j.matdes.2006.12.016.

Pidaparti, R.M.V. and Palakal, M.J. (1993). Material model for composites using neural networks. *AIAA Journal*, 31(8), 1533–1535. https://doi.org/10.2514/3.11810.

Ramadan Suleiman, A. and Nehdi, M. (2017). Modeling self-healing of concrete using hybrid genetic algorithm–artificial neural network. *Materials*, 10(2), 135. https://doi.org/10.3390/ma10020135.

Riffat, S.B. and Qiu, G. (2013). A review of state-of-the-art aerogel applications in buildings. *International Journal of Low-Carbon Technologies*, 8(1), 1–6. https://doi.org/10.1093/ijlct/cts001.

Saad, S., Alaloul, W.S., Ammad, S., Qureshi, A.H., Altaf, M. and Rasheed, K. (2020). Design Phase Carbon Emission Prediction Using a Visual Programming Technique. *2020 2nd International Sustainability and Resilience Conference: Technology and Innovation in Building Designs*. https://doi.org/10.1109/IEEECONF51154.2020.9319978.

Saad, S., Salah Alaloul, W., Ammad, S., Hannan Qureshi, A., Mohsen Mohammed Alawag, A., Kumar Oad, V. and Altaf, M. (2020). A noded carbon emission tool (NCET) to measure embodied carbon (EC) in compositional cementitious materials. *Solid State Technology*, 63(6), 4099–4105.

Saad, S., Alaloul, W.S. and Ammad, S. (2022). Role of cyber-physical systems in smart cities. *In*: D.W.S. Alaloul (Ed.), *Cyber-Physical Systems in the Construction Sector* (1st Edn.), p. 19. CRC Press. https://doi.org/9781003190134.

Shabani, M.O. and Mazahery, A. (2012). Application of finite element model and artificial neural network in characterization of Al matrix nanocomposites using various training algorithms. *Metallurgical and Materials Transactions A*, 43(6), 2158–2165. https://doi.org/10.1007/s11661-011-1040-1.

Sivasankaran, S., Narayanasamy, R., Ramesh, T. and Prabhakar, M. (2009). Analysis of workability behavior of Al–SiC P/M composites using backpropagation neural network model and statistical technique. *Computational Materials Science*, 47(1), 46–59. https://doi.org/10.1016/j.commatsci.2009.06.013.

Takahashi, K. and Tanaka, Y. (2016). Material synthesis and design from first principle calculations and machine learning. *Computational Materials Science*, 112, 364–367. https://doi.org/10.1016/j.commatsci.2015.11.013.

Trebar, M., Susteric, Z. and Lotric, U. (2007). Predicting mechanical properties of elastomers with neural networks. *Polymer*, 48(18), 5340–5347. https://doi.org/10.1016/j.polymer.2007.07.030.

Vijayaraghavan, V., Garg, A., Wong, C.H., Tai, K., Singru, P.M., Gao, L. and Sangwan, K.S. (2014). A molecular dynamics based artificial intelligence approach for characterizing thermal transport in nanoscale material. *Thermochimica Acta*, 594, 39–49. https://doi.org/10.1016/j.tca.2014.08.029.

Wang, X., Chen, A. and Liu, Y. (2022). Explainable ensemble learning model for predicting steel section-concrete bond strength. *Construction and Building Materials*, 356, 129239. https://doi.org/10.1016/j.conbuildmat.2022.129239.

Xiao, Q., Li, C., Lei, S., Han, X., Chen, Q., Qiu, Z. and Sun, B. (2021). Using hybrid Artificial Intelligence approaches to predict the fracture energy of concrete beams. *Advances in Civil Engineering*, 2021, 1–12. https://doi.org/10.1155/2021/6663767.

CHAPTER

4

Industry 4.0 and Construction

Kumeel Rasheed, Syed Saad, Syed Ammad and Muhammad Tariq Bashir

Introduction

The Fourth Industrial Revolution, more frequently known as Industry 4.0, is the combination of cutting-edge digital technology into diverse industries, including manufacturing, supply chain, and construction. The construction industry is undergoing a paradigm shift as an outcome of the rapid growth of AI, which is revolutionizing the method in the manner materials are formed, processed, and adopted in the construction industry. This transition is being enhanced by AI in material science, which is providing previously unheard-of prospects for increased efficacy, sustainability, and affordability. The upsurge of Industry 4.0 has positioned the construction sector on the verge of a technological revolution. Construction has a lot of prospects for transformation, which is considered by the synthesis of digital and artificial intelligence (AI) technology. The use of materials in building is being revolutionized by AI, which is making substantial progress in a number of crucial fields including material science. For greater qualities, increased durability, and sustainability, the developing materials make use of advances in nanotechnology, biotechnology, and sophisticated production processes, such as, composite products, nanomaterials, and elements that heal themselves are a few noteworthy examples (Salem, 2023). Composite materials, particularly fiber-reinforced polymers (FRPs), oriented strand board (OSB), shape memory polymers (SMP) are exceedingly preferred when a need for lightweight materials with exceptional durability due to their significant strength-to-weight ratios. Few of the main aspects that these materials possess are exceptional resistance against corrosion, fatigue, and extreme temperatures, that make it suitable for various building applications including road construction, aviation, and wide-range construction projects. In the construction sector, Industry 4.0 concepts find practical applications. A good illustration is the effective use of Virtual Reality (VR) in the product of construction machines, which leads to shorter product cycles, better quality, and reduced prices. Building information modeling (BIM), Virtual Reality (VR), Augmented Reality (AR), Internet of Things (IoT), Big Data, and Cyber-Physical Systems (CPS) are just a many of the industry 4.0 related technologies that the construction sector is rapidly embracing. Due to their broad relinquishment, BIM, cloud, mobile, and modularization technologies have reached a relative position of maturity. Advanced robotics and automation, together with prefabrication and modular building techniques, allows to decrease resource inefficiencies, construction time, and material waste (Wu et al., 2021). It also results in safer job sites, more efficient project execution, and more environmentally friendly building methods.

With the deployment of AI and industry 4.0 technologies, the construction sector, which is renowned for its labor-intensive and lengthy procedures, is poised for upheaval. The development of slice edge materials and digital technology has led to an enhancement in the effectiveness, cost, and sustainability of construction styles. Businesses are becoming more and more aware of the enormous potential that technology presents for the construction sector. They may efficiently decrease waste, minimize effort duplication, and improve control over project quality, time, and budgets by utilizing digital tools and procedures. The digitization of building processes also makes it possible for businesses to collect and analyze enormous volumes of data. Circular economy concepts could potentially be used in the construction sector more easily with the help of Industry 4.0 technology as materials and components can possibly be tracked, recorded, and recycled using digital tracking and traceability systems at the end of their lives. To assess and optimize materials for improved performance, robustness, and ecological consequences, this revolution's operation of AI in material wisdom is essential. The creation of new materials with advanced characteristics is one of the major developments in material wisdom fueled by AI. For example, nanotechnology has produced materials with exceptional strength, inflexibility, and conductivity. By using these materials, lesser material is consumed during construction, and energy effectiveness is increased. Also, new materials are being discovered and their implicit uses are being eased by AI algorithms. Experimenters can find materials with certain features, such as thermal resistance, fire resistance, or aural parcels, designed to fulfill the requirements of colorful structure systems through data analysis and computational simulations. Construction experts can upgrade building layouts, material alternatives, and power systems to attain better degrees of sustainability by applying computerized simulations and study (Abanda & Byers, 2016). As a result, there is a decrease in energy use, greenhouse gas emissions, and an improvement in quality of the indoor environment. Nowadays, companies are becoming increasingly aware of the enormous potential that technology presents for the construction sector. They may efficiently decrease waste, minimize effort duplication, and improve control over project quality, time, and budgets by utilizing digital tools and procedures. The digitization of building processes also makes it possible for businesses to collect and analyze enormous volumes of data as this data-driven strategy offers insightful information that can be used to make decisions and promote proactive risk reduction. The use of AI in the construction industry is still in its nascent stages, but it has the potential to revise the way that structures are designed, erected, and operated. AI is making it possible to make more effective, sustainable, and safer structures as this will have a bigger influence on the construction sector as it develops. Here are some more facts concerning the application of AI in material science:

- AI is being used to produce new self-healing materials, which might fully change how we fix damaged structures.
- To produce new materials that can absorb carbon dioxide from the atmosphere and possibly decelerate down climate change, AI is being applied.
- With the help of AI, new materials are being created that may be utilized to make new kinds of sensors and actuators, which might affect in advancements in robotics and other industries.
- Machine learning (ML) algorithms are used to forecast the behavior and performance of materials over time, enabling preventive maintenance methods and extending the life of constructions.
- The integration of AI in material wisdom is fostering collaboration and knowledge sharing within the construction industry as experimenters, architects, and material manufacturers can work AI platforms and databases to change perceptivity, conduct common exploration, and drive invention in materials and construction practices.
- AI-powered virtual testing environments allow for immediate material prototyping and modeling, which eliminates the need for lengthy physical testing and quickens the creation of new materials.

Overview of Industry 4.0

Industry 4.0, the Fourth Industrial Revolution, is the incorporation of improved digital technology and automation across a variation of industries, involving manufacturing, supply chain operation, and services. The confluence of technologies like AI, cloud computing, IoT, CPS, robotics, big data analytics, and cumulative manufacturing is what it represents. It denotes a profound conversion in how productions are conceptualized, produced, and delivered. The core idea behind Industry 4.0 is to produce "intelligent industries" and "smart systems" that connect devices, processes, and people in order to enable real-time data interchange and decision-making (Roblek et al., 2016). Industry 4.0's effects go beyond particular factories or procedures. It entails developing digital supply chains, enhancing distribution and logistics, and redefining business paradigms. The unified collaboration and integration across many industries and stakeholders promotes innovation, efficiency improvements, and new openings for profitable growth. Industry 4.0, as a whole, represents a paradigm change in industrial operations, utilizing revolutionary technology to promote digital transformation, boost productivity, and build sustainable and flexible industries in the face of ever-changing market demands and global issues.

Core Principles and Objectives of Industry 4.0

The fundamental principles of Industry 4.0 work together to shift many sectors towards more connection, digitization, and automation. By adopting these concepts, organizations increase productivity, boost sustainability, and acquire a competitive advantage in the Fourth Industrial Revolution's fast changing environment. The key principles are:

Boosted Operational Efficiency

By utilizing new technologies, Industry 4.0 seeks to increase productivity and operational effectiveness. Businesses can optimize resource use, minimize waste, and simplify their operations using computerization, real-time data analysis, and efficient algorithms. They can also accomplish flexible methods of production, facilitating adaptation, effective prototype design, and just-in-time manufacturing, by integrating technology like cyber-physical systems and IoT.

Improved Quality and Durability

The basic premise of Industry 4.0 is to improve product durability and quality through professional monitoring, management, and preventative maintenance techniques. Organizations can recognize and take care of possible complications before it results in flaws or malfunctions in equipment by utilizing sensors, information analysis, and ML.

Enhanced Innovation

In the framework of Industry 4.0, improved innovation is accomplished via the incorporation and application of the latest developments like AI, ML, and expert robotics. These technologies give organizations the ability to simplify and automate their operations, boosting productivity, efficiency, and velocity in product development and processing. Advanced robotics systems also offer automation and accuracy in manufacturing, minimizing mistakes, enhancing quality, and speeding up production (Stavropoulos et al., 2018). Companies may acquire an advantage over their rivals and set themselves apart in their field by adopting digital transformation and embedding these modern technologies into their day-to-day operations.

Sustainable Resource Management

In many aspects, Industry 4.0 is essential for advancing sustainable resource efficiency. First, organizations may monitor and optimize energy use by deploying smart systems to manage energy and using IoT-enabled devices, which significantly lowers wasteful usage of energy and carbon emissions (Saad et al.,

2020). Second, the application of AI and ML algorithms can assist in locating opportunities for energy optimization and offer energy-efficient substitutes.

Optimized Safety Protocols

The combined application of technology and automation is a key component of Industry 4.0's priority on improving workers' safety. Businesses could reduce the necessity for human participation in hazardous processes and lower the likelihood of injuries and crashes by implementing robots and automated systems. Robotic systems with next-generation sensors and AI algorithms are capable of carrying out jobs that are hazardous or physically taxing for humans (Rasheed, Saad et al., 2021). The places the robots may work in do not have to be dangerous for human employees, whether they contain high heat, poisonous materials, or other hazardous circumstances. Technologies for virtual reality (VR) and augmented reality (AR) also contribute to increased safety. Corporations can educate their workers on safety regulations and emergency protocols without putting them in danger by offering virtual training environments. AR could be used to add safety instructions or alerts on actual machinery, direct workers through challenging jobs, or detect possible dangers. By enabling people to concentrate on more difficult and valuable jobs, this automation not just enhances safety but also maximizes efficiency and productivity.

Evolution of Industrial Revolutions: From Industry 1.0 to Industry 4.0

The development of industrial revolutions represents how human civilization has changed from being dependent on manual labor to the digital era. It includes a number of significant milestones in the development of industry and technology. Every industrial revolution led to significant changes in production techniques, manufacturing procedures, and the wider organizational structure of enterprises as shown in Figure 4.1.

First Industrial Revolution (Industry 1.0)

The late 18th century saw the dawn of the First Industrial Revolution, often referred as Industry 1.0, which represents a crucial turning point in the evolution of humanity. It was a time of significant transformation marked by the shift from an agrarian to a mechanized manufacturing and steam power-dependent civilization. During Industry 1.0, new tools and technology, specifically the steam engine, progressively took the position of manual labor. By offering a dependable and effective source of electricity, James Watt's innovation revolutionized several sectors. The mechanization of production procedures was made possible by the introduction of steam engines to power factories and machines. Large-scale factories replaced decentralized village industries as a result of the invention of steam power. Previously, miniature workshops

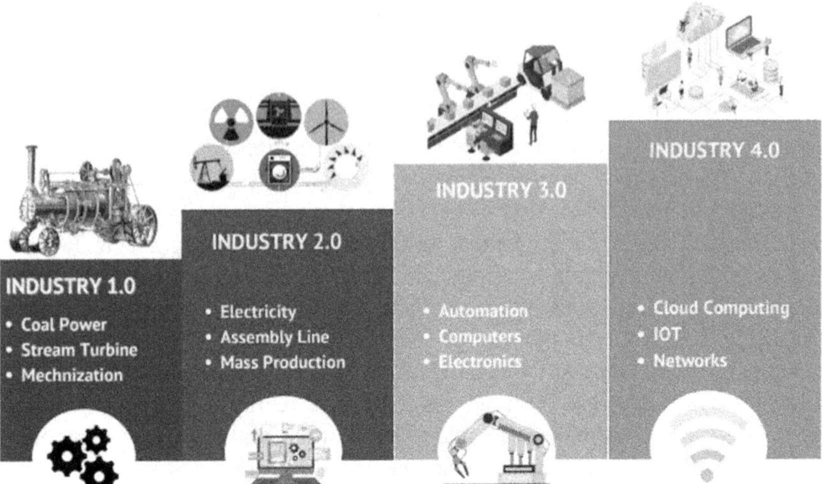

Figure 4.1: Industrial revolution.

or solitary artisans manufactured the majority of the items. But as factories emerged, manufacturing was concentrated in specialized spaces where robots could do jobs more rapidly and effectively than humans.

The social and economic ramifications of Industry 1.0 were significant. The invention of steam-powered locomotives and steam vessels, which allowed the transportation of individuals and commodities across great distances, was one of the important developments that resulted from it (Sachs, 2019). As people moved from rural regions to cities in pursuit of factory jobs, the development of industry and the spread of railways accelerated urbanization. By proving the potential of mechanization and the transformational power of technology in promoting economic progress, the First Industrial Revolution established the groundwork for succeeding industrial revolutions. It prepared the foundation for the succeeding waves of industrialization that would go on to influence the course of history for millennia to come.

Second Industrial Revolution (Industry 2.0)

The Second Industrial Revolution, frequently known as Industry 2.0, occurred in the late 19th and early 20th centuries and led to major developments in manufacturing and technology. The use of electricity as a commonly used energy source was one of the significant advancements during Industry 2.0. Due to the increased flexibility, efficiency, and controllability, electric motors were capable of replacing steam engines in manufacturing. Industrial processes were transformed by the accessibility of electricity, which resulted in higher productivity and the capacity to run machines more precisely and automatically. Electricity's invention, the progression of mass manufacturing techniques, and the birth of the assembly line were all defining factors of this time.

The mass manufacturing techniques were implemented, which was a significant Industry 2.0 breakthrough as well. Eli Whitney's invention of interchangeable components and Henry Ford's popularization of the assembly line, respectively, made it possible to produce items on a wide scale with efficiency. In factories, manufacturing was streamlined to save costs and boost productivity by segmenting jobs into smaller, more specialized processes and using standardized parts (Dixit et al., 2017). The development of the telegraph and telephone enabled businesses to enhance operational coordination and streamline information communication. These advancements facilitated quick and long-distance correspondence, leading to more effective coordination of operations.

Simultaneously, transportation underwent a significant transformation with the advent of railways and the invention of vehicles. This revolutionized the movement of people and goods, enabling faster and more reliable transportation methods. Many different economic sectors were significantly influenced by Industry 2.0, with areas such as telecommunications, transportation, and manufacturing undergoing radical changes. Results of this transformational period included increased productivity, economic growth, and the introduction of modern consumer goods. The foundation for the modern industrial environment was laid and the stage for succeeding industrial revolutions was created by the revolutionary inventions of Industry 2.0, which accelerated technical advancement.

Third Industrial Revolution (Industry 3.0)

A significant step in manufacturing and mechanization happed during the late 20th century with the arrival of the Third Industrial Revolution, also known as Industry 3.0. During this period, we saw the development of computers and digital technology as well as the machine-driven production process of different types. The wide use of computer systems and other digital technologies in industrial situations was one of Industry 3.0's major developments. Computers have made it feasible to streamline formerly manual operations, which has enhanced efficiency, accuracy, and production. Microprocessors have become more widely available and less expensive, and electronic components have shrunk in size, giving organizations the potential to incorporate technology gadgets into their processes. Industrial robots were introduced to expedite and accurately perform repetitive tasks, resulting in increased efficiency. These robots were capable of tasks such as assembly, welding, and material handling, reducing the reliance on human labor and enhancing productivity.

The implementation of information technology systems extended beyond manufacturing to other areas such as supply chain logistics, inventory management, and quality assurance, streamlining operations and improving efficiency (Jiang et al., 2022). The integration of automation and computer technologies in Industry 3.0 brought about advancements in manufacturing processes. Just-in-time manufacturing techniques became feasible, ensuring supplies were delivered precisely when needed, thereby reducing waste and inventory costs. Computer-aided design (CAD) and computer-aided manufacturing (CAM) have brought about a revolution in product design and manufacturing, facilitating the creation of superior-quality and customized items. It has ushered in a new era of product development, where innovation and customization are at the forefront, resulting in higher-quality and uniquely configured products. The Industry 3.0 triggered a paradigm change in the industrial atmosphere, resulting in more productivity, lower costs, and better product quality. The development of machines and digital technology, during this time period laid the groundwork for the subsequent industrial revolution, known as Industry 4.0, in which cyber-physical systems and networked technologies would continue to change the manufacturing industry.

Fourth Industrial Revolution (Industry 4.0)

The Fourth Industrial Revolution, generally known as Industry 4.0, is the latest stage in the evolution of product and technology. It establishes a new era of connection, digitization, and technological advancement on top of the root set by earlier industrial revolutions. Industry 4.0 is considered by the merger of the IoTs, data analytics, AI, CPSs, and other revolutionary technologies into the production developments as this involves the digitalization of tangible assets, the utilization of smart sensors and gadgets, and the efficient transaction of data between computers, systems, and people. Interconnectedness is one of the core aspects behind Industry 4.0. The IoT hyperlinks and connects machines and devices, permitting for real-time data exchange and decision-making. The enhancement of industrial procedures, protective maintenance, and the growth of intelligent, self-learning systems are all made possible by this connectedness. Data is essential to Industry 4.0, by the use of powerful analytics and AI algorithms to draw understandings, spot trends, and make educated decisions.

Key Technologies Driving Industry 4.0

A fusion of revolutionary technologies that are changing the industrial environment has accelerated the emergence of Industry 4.0. The unparalleled stages of connection, automation, and data-driven decision-making are being empowered by these important technologies. The IoT, which allows a diverse device to interconnect and share data via the internet, is one of the key technologies. Due to the real-time monitoring, gathering data, and analysis made possible by this interconnection, several industries have experienced increased productivity and predictive maintenance (Jiang et al., 2022). Along with IoT, ML and AI are transforming the ways in which data is processed and analyzed. These innovations enable systems to recognize patterns in data, take independent actions, and streamline procedures. By using predictive analytics and pattern recognition, systems which use AI can develop manufacture effectiveness, product quality, and supply chain management. The additive manufacturing, can also be referred to as 3D printing, which makes it possible to produce complicated and unique products which uses less waste and faster turnaround times as the technology has the massive potential to transform conventional manufacturing developments through rapid prototyping and local production.

Internet of Things (IoT) and Its Role in Industry 4.0

In the Industry 4.0, the IoT is a key technology pillar that is transforming how industries function and communicate with their surroundings. IoT is described as a network of interconnected devices, sensors and machines that share information and communicate on the basis of internet without direct human development. IoT primarily makes it possible to develop 'smart' systems and procedures as sensors can be added to factories, machinery, vehicles, and even common objects, which can then be connected to IoT systems (Rasheed et al., 2022). This makes it possible for them to produce and communicate data about their performance, status, and environmental factors. With the use of this data, businesses may remotely monitor operations, spot inefficiencies, and take preventative maintenance measures that will boost operational effectiveness and decrease downtime.

The practice of predictive maintenance has been affected by IoT in Industry 4.0. Manufacturers can forecast when a component is likely to break or need maintenance by gathering real-time data from the equipment. With this proactive strategy, unplanned downtime is reduced, the lifespan of machinery is increased, and maintenance plans are optimized. Additionally, IoT-driven data analytics provide insights into industrial processes, enabling ongoing optimization. Businesses can pinpoint workflow bottlenecks, improve product quality, and detect bottlenecks by evaluating data from various phases of the production process. However, there are drawbacks to IoT integration in Industry 4.0, such as worries about data security and privacy. Strong cybersecurity measures are therefore necessary to safeguard sensitive data and guarantee the reliability of networked systems.

Artificial Intelligence and Machine Learning in Industry 4.0

AI and ML are driving Industry 4.0, which is regarded as transforming industries by providing them with previously unreachable levels of automation, intelligence, and data-driven choices. AI is considered as the imitation of human intelligence in machines because it allows them to continue the operations that are normally done by human cognition. A branch of AI is known as ML, which gives computers the ability to recognize patterns in data, increase their functionality, and make decisions on their own. AI and ML are essential to the implementation of Industry 4.0 since they transform many aspects of production and business operations. Predictive maintenance is one of the key uses of AI and ML in Industry 4.0. AI systems can predict when equipment is likely to fail by evaluating real-time data from sensors and devices, enabling manufacturers to do maintenance before failures happen (Abdul Hannan Qureshi et al., 2020). With this strategy, unplanned downtime is reduced, equipment uptime is increased, and maintenance expenses are reduced.

AI-driven insights also make it possible to continuously enhance industrial procedures. Huge datasets are analyzed by AI to spot inefficiencies, spot trends, and suggest improvement tactics. AI-enabled smart manufacturing systems are able to modify production processes in real-time based on data inputs. For example, when working with fluctuating input materials, a production line, can automatically readjust parameters to maintain constant product quality. This iterative method improves product quality overall, lowers faults, and boosts manufacturing efficiency. Similarly, AI-powered design tools assist engineers in developing creative and efficient designs for products. ML algorithms examine simulations and historical data to produce ideas that fulfill predetermined criteria, fostering innovation and speeding up the design process (Qureshi et al., 2020).

Big Data Analytics and Its Significance in Industry 4.0

Industry 4.0 primarily relies on Big Data Analytics to process, analyze, and derive useful insights from enormous information produced by linked devices, equipment, and systems. The ability to uncover hidden patterns, correlations, and trends in the data holds transformative potential for various industries. Big Data Analytics is a cornerstone of Industry 4.0 since traditional analytic techniques might overlook significant trends in the age of data abundance. Manufacturing companies can better comprehend their operations by utilizing the data flood from sensors and manufacturing processes, which promotes efficiency and creativity. It can recognize inefficiencies, identify the underlying causes of issues, and make data-driven modifications for improvement (Meier et al., 2023). These data analytics can conveniently be used to monitor and implement alterations in real time with respect to quality control. To guarantee constant product quality, data from sensors and production systems are regularly examined. This is especially important in fields like aerospace and medical equipment where even minute changes can cause flaws or failures.

Cloud Computing and Its Impact on Industry 4.0

In the light of Industry 4.0, cloud computing is an innovative force that is transforming how industries store and process data, interact, and improve their operations. The large volume of data created by linked devices and systems is one of the primary outcomes of cloud computing in Industry 4.0. Real-time processing, interpretation, and storage are made possible by cloud platforms, which offer the infrastructure and storage capacity required to handle this abundance of data. Industries often encounter changes in data demand as a result of seasonal patterns or unexpected spikes in usage. With the ability to flexibly scale up or down in response to demand, cloud solutions eliminate the need for over-provisioning and resource waste. This is especially useful in sectors like e-commerce or manufacturing where agility and responsiveness are critical. Cloud computing also improves data sharing and collaboration (Naeem et al., 2022), in which cloud-based solutions provide smooth data sharing in an Industry

4.0 environment where many stakeholders along the supply chain need access to real-time data. Cloud Computing helps Industry 4.0 in reducing the cost of it, as typical IT infrastructure requires significant up-front expenditures for hardware and upkeep. One of the models known as Pay-as-you-go are available with cloud solutions, in which companies only pay for the resources they utilize. Smaller businesses are now able to access computer capabilities that were previously out of their price range, developing innovation and competition. The effect of cloud computing will increase as Industry 4.0 advances further, empowering industries to fully utilize linked technology for improved operations and long-term growth.

Impact of Industry 4.0 on Construction

Industry 4.0 has a significant influence on the construction sector, bringing in a new era of productivity, sustainability, and innovation. Through the use of advanced digital technology, this technological revolution is redefining the construction industry by transforming old procedures, facilitating cooperation, and enhancing project outcomes. The alteration of building procedures and workflows is one of the key effects. Industry 4.0 technologies simplify stakeholder coordination, communication, and project administration. Construction experts may view designs in a virtual environment before execution, minimizing mistakes and optimizing resource allocation, especially to Building Information Modeling (BIM) and digital platforms (Saad et al., 2021). In terms of Industry 4.0's effects on the construction industry, digital twin technology is a game-changer. By building a digital duplicate of a real structure, it makes it possible to simulate, analyze, and monitor situations in real time. Early design defect detection, effective project management, and continual monitoring for upkeep and operating needs are all made possible by this technology. Digital platforms and technologies made possible by Industry 4.0 allow for the building of thorough digital models of construction projects. These models, which frequently utilize BIM, include information on architecture, structures, and systems. Before construction starts, stakeholders may see the project in great detail owing to digital models, helping them spot possible conflicts or design problems early on. This improves communication between architects, engineers, contractors, and other stakeholders in addition to reducing expensive rework. The exchange of real-time data and the centralization of project information promote effective decision-making, greatly lowering project timeframes and raising overall project quality.

Optimization of Construction Sequencing and Scheduling

Construction project sequencing and schedule optimization has become a key component of Industry 4.0, leveraging new technologies to completely transform how construction projects are organized, carried out, and managed. Advanced algorithms and data-driven methodologies developed by Industry 4.0 enable construction planners to develop dynamic timetables (Menezes et al., 2019).

Task dependencies, resource availability, and external factors like weather conditions are all taken into account by these schedules. These algorithms can figure out the ideal job sequencing that avoids bottlenecks, cuts down on idle time, and assures effective resource usage using AI-driven analysis. Construction scheduling gains a new dimension as real-time monitoring made possible by the IoT. Sensors built into construction tools, supplies, and even workers gather information on how jobs are going. This information is transmitted to a central system, where it is immediately examined. This feature enables project managers to quickly identify plan deviations and make corrections, minimizing delays from cascading across the project. Optimized scheduling and sequencing have effects that go beyond time savings. It has an immediate impact on project quality, cost control, and resource management. Overtime, rework, and idle time expenses may be reduced by making sure that resources are distributed effectively. A well-optimized timeline also lessens the possibility of resource disputes or shortages, which facilitates a smoother building process. Due to unanticipated situations, the construction sector has a history of experiencing delays. However, Industry 4.0's sequencing and scheduling optimization helps to reduce these risks. Algorithms can provide probabilistic insights into anticipated delays by taking into account different scenarios and making use of past data. Project managers can proactively allocate resources and modify timelines as a result of this insight.

Prefabrication and Modular Construction in Industry 4.0

Prefabrication in the context of Industry 4.0 refers to the fabrication of building components in controlled environments away from the construction site. This covers building modules in their whole as well as structural components, wall panels, flooring systems, and more. Accurate and thorough component designs are being produced with the aid of cutting-edge digital design technologies. The components are subsequently manufactured with extreme accuracy and quality because of computer-controlled technology, which translates these designs into exact instructions (Menezes et al., 2019). Similar to prefabrication, modular construction entails assembling these prefabricated parts on the job site like building bricks. By streamlining the process, less on-site labor is needed, less time is spent building, and less material is wasted. The incorporation of IoT devices makes modular building more efficient in the context of Industry 4.0. These tools keep an eye on how prefabricated parts are made, making sure that quality requirements are followed. Additionally, they provide real-time tracking throughout assembly and transit, improving quality assurance and traceability. Under Industry 4.0, prefabrication and modular construction have a wide range of effects which are:

- Firstly, the techniques significantly shorten project duration. Time-consuming on-site operations are reduced by producing components off-site while simultaneously prepping the building site. This is especially useful for tasks with constrained timelines or inclement weather.

- Secondly, the accuracy attained by robotic production and digital design raises the caliber of building. By manufacturing components in a regulated system, environmental mistakes are reduced such as higher structural integrity, greater energy efficiency, and generally improved building performance are all impacted by this precision.
- Thirdly, prefabrication and modular construction are important aspects of sustainability. The controlled manufacturing environment reduces material waste, and the assembly process generates fewer site-related disruptions. Also, Industry 4.0's sustainability objectives are in line with the effective use of resources and lower energy usage.

Cyber-Physical Systems (CPS) in Construction

By combining the physical and digital realms into one linked ecosystem, CPS are at the leading edge of Industry 4.0's technology integration and changing the construction sector. CPS are a combination of physical elements, sensor networks, data analytics, and computational algorithms. This convergence gives construction operations and procedures a level of efficiency, precision, and adaptability never before possible. CPS plays a significant and complex role in the transformation of construction workflows and processes. The main component of CPS is the integration of real-time data collection, processing, and usage into construction operations (Saad, Alaloul, Rasheed et al., 2022). Many different types of data, from environmental factors to structural performance indicators, are gathered through sensors embedded into building tools, equipment, and materials. Advanced algorithms are then used to process this data, allowing quick decision-making, the foretelling of prospective problems, and the optimization of resource allocation. CPS manages automated activities and increases machinery efficiency by using advanced algorithms to analyze real-time data. Consider the carefully regulated concrete pouring that CPS directs in response to variables like temperature and humidity to ensure ideal recovery, as with this automation, CPS increases accuracy while reducing labor-intensive tasks and adding a new level of efficiency to the procedure. CPS acts as a steady sentinel throughout the life of a project, adjusting activities in response to changing site circumstances or shifts in material supply (Saad, Alaloul and Ammad, 2022).

CPS in Equipment and Machinery Automation in Construction

The revolutionary impact of CPS on the automation of equipment and machinery highlights a significant change in construction processes. These massive objects change into clever data collectors by adding sensors to the construction equipment. By observing variables like vibrations, temperatures, and wear indicators, these machines can monitor their performance. They can inform the appropriate people when they notice something odd so that they can address it before it becomes a significant issue (Ammad et al., 2022). The data arrangement is composed using a constant stream of real-time data that includes performance parameters, fuel usage, and operational health. Advanced algorithms combine this data to provide insights that may be put into practice, promoting the efficient use of equipment, reducing downtime, and coordinating resource allocation.

Digital Twin Technology

The development of digital twin technology is a game-changing factor in the dynamic world of contemporary construction, fundamentally altering the course of building project lifecycles. A digital twin is essentially a virtual replica of the real thing. Making a model is only one aspect of this; more accurately, it involves bringing the real thing to life in the digital world. This digital replica receives data from sensors and other sources, providing a detailed picture of how the actual asset functions. Construction could utilize digital twins in a variety of ways. They aid in the first stages of design, enabling architects and engineers to refine their ideas before to the commencement of real building. As soon as building starts, data from the site is uploaded to the virtual twin, producing a digital record of the project's development. This helps in keeping project managers informed and enables them to make wiser choices when problems do arise. In order to streamline collaboration and reduce misunderstandings, digital twins provide a common platform for insights that encourage good communication among team members (Teng et al., 2021).

Improved Collaboration and Communication in Construction Projects

Beyond the adoption of new technologies, Industry 4.0 has a significant impact on the construction industry since it is transforming how teams interact and cooperate on projects. Platforms for cooperation in the cloud are the foundation of this change. These platforms offer a digital environment where any team member may access project-related data and documents and participate as this eliminates geographical barriers and makes it possible for teams that are spread out to work together seamlessly. To keep everyone informed and on the same page, design blueprints, progress reports, and other project necessities can be updated in real-time. Similarly, Technology advancements in VR and AR are giving collaboration new meaning (Cipresso et al., 2018). Despite being far apart, these technologies enable team members to virtually engage themselves in the project environment. A virtual building site may be toured by architects, engineers, and other interested parties who can then analyze designs and spot any problems. Clearer communication and speedier decision-making are facilitated by this immersive experience. Real-time data exchange through a variety of digital channels is made possible by Industry 4.0. Project management software, video conferencing, and instant messaging can all be utilized to keep team members in communication. No matter where team members are physically located, communication is optimized to guarantee that project updates, questions, and decisions are made without delay.

Challenges and Opportunities for AI in Construction

The application of AI in the construction industry creates a landscape filled with prospects and difficulties. The technology has the potential to completely revolutionize the different areas of construction building, however the availability and quality of data creates main hurdles in the deployment of AI in the construction industry. Large volumes of data are produced during construction projects, but they are frequently dispersed throughout several systems, managed by multiple stakeholders, and in different formats. An exhausting task may be integrating and standardizing this data for AI-driven insights. Furthermore, for AI algorithms to provide useful outputs, the correctness and dependability of the data must be guaranteed. It takes effective data management methods and technology to overcome these data-related difficulties. Data is produced in great quantities during construction projects, including design blueprints, material specifications, progress reports, and equipment utilization (Huang et al., 2018). However, these information silos, such as project management software, BIM systems, and finance databases, often keep these data separately. Strong data collecting and management procedures are required to assemble and arrange these many data sources into a logical framework for AI study. In applications using AI, data dependability and accuracy are of the utmost significance. Incomplete or inaccurate data might provide biased conclusions and incorrect forecasts. This could lead to poor judgments in the context of construction that have an impact on project finances, deadlines, and success in general. Validating data and preserving its integrity throughout the course of its lifespan are both necessary for ensuring data correctness (Duggineni, 2023). Similarly, seamless interchange of data across various systems, or interoperability, is a complex issue in the construction industry. Different stakeholders use a range of software tools, including architects, engineers, contractors, and subcontractors. It is a challenging endeavor to integrate these instruments so that data may flow easily across different stages of a building project. The ability of AI to produce comprehensive insights hinges on removing these data restrictions and fostering effective teamwork.

Workforce Adaptation and Skill Requirements for AI Implementation

The application of AI in the construction sector highlights the importance for skill growth and workforce adaptability. To completely utilize AI systems, the technological progression requires a workforce that not only understands their complexities but also works well with them. The shift to a construction sector enhanced by AI highlights the necessity of reskilling and upskilling the current workforce. The merger of technical competence and AI literacy will now be necessary for traditional building professions. It's important for professionals to become familiar with the ideas behind AI, data analytics, and AI-powered solutions. This entails developing customized training programs that enable the workforce to use AI for better project results and decision-making. The key to effective AI integration in construction is human-AI system collaboration. It is critical to see AI as a partner rather than a rival. By providing data-driven insights, AI enhances human capabilities and gives professionals more time for strategic thinking and creative problem-solving.

Legal and Ethical Considerations in AI Adoption

The use of AI in design processes has complicated issues with ownership of designs and intellectual property rights. The distinction between human and machine creativity can be dissolved by AI systems' ability to produce inventive designs on their own. The question of who should be legally responsible for ownership arises: should it be the AI's creator, the designer who selects the ideas it generates, or the AI itself? A legal framework that specifies ownership rights and obligations in this innovative AI-design environment is required to address this difficulty. Although effective, AI systems may unintentionally reinforce biases found in the data they are trained on, this also applies to the construction industry, as biases can be seen in selections made about resource allocation, project management, and risk assessment (Akter et al., 2021). Rigorous testing, ongoing supervision, and algorithmic fine-tuning are all necessary to ensure fairness in AI-driven building systems and to minimize any biased outcomes.

Economic Impacts and Cost-effectiveness of AI in Construction

The cost effectiveness is one of the pillars of AI adoption in the building industry. The advantages AI delivers are clearly illustrated by a careful review of the financial inflows and outflows connected with AI adoption. The feasibility of AI integration in relation to the added economic value is determined by stakeholders through careful examination. AI technologies are channels for significant cost reductions and increased efficiency in the building operations. AI-driven insights provide an immediate financial benefit by maximizing resource allocation, project scheduling, and material utilization (Pan & Zhang, 2023). Significant cost-effectiveness and increased productivity across projects are possible with simplified procedures and reduced waste. The economic benefits of AI go beyond short-term cost reductions as consistent efficiency gains, less project risks, and improved project results all lead to long-term financial gains. Over the course of a building project, these cumulative benefits help the bottom line. Analyzing these long-term financial gains reveals AI's contribution to the transformation of construction into a financially responsible industry.

Industry 4.0 in Construction: Key Technologies

Construction will see a variety of revolutionary advantages as a result of the introduction of Industry 4.0 concepts and key technologies. These advantages will elevate current practices and establish new benchmarks for productivity, teamwork, and sustainability. The capabilities and results of the construction business are amplified by these essential technologies when used together:

Artificial Intelligence and Machine Learning in Construction

AI algorithms estimate project hazards, timetable alterations, and resource requirements by examining past and current data. Due to the ability to take proactive decisions and reduce risks, project managers are better equipped to handle problems before they get worse. AI-powered algorithms examine the building process and look for any flaws or design requirements that have been strayed from. These technologies steer clear of future costly corrections by seeing problems in real time. AI-driven quality control boosts project productivity by reducing rework while simultaneously ensuring that projects are completed in accordance with requirements. ML algorithms examine a wide range of factors, gain knowledge from the past, and spot patterns that people would overlook. Accurate risk assessments that guide resource allocation, project scheduling, and risk mitigation tactics are the result of this process.

Robotics and Automation in Construction Processes

Modernizing building processes has advanced significantly with the use of robots and automation, including the use of drones. To redefine the dynamics of building sites, robotic devices take front stage. These robotic assistants do a wide range of formerly labor-intensive jobs, such as pouring concrete and placing bricks. Their accuracy and consistency speed up project schedules, reduce human error, and ultimately improve project quality and efficiency. Equipment used in material handling and construction are also affected by automation. Construction sites may be effectively managed and transported by automated technologies. In addition to shortening project durations, this logistics optimization reduces resource inefficiency and waste. Drones and autonomous cars are two prominent examples of how technology is affecting the construction industry. Drones fly over building projects and take overhead pictures that may be used for site analysis and progress tracking. Using autonomous vehicles, building sites can be surveyed and materials may be transported. Collectively, these technologies transform data collecting, improve security, and quicken project management.

Internet of Things and Sensor Technologies in Construction

In building projects, smart sensors are becoming essential instruments for monitoring the structural health. These sensors gather information on structural conditions continually and can spot fractures, irregularities, or deviations immediately (Rasheed, Shahzad et al., 2021). Smart sensors improve the safety and longevity of building projects by seeing possible problems early and enabling proactive maintenance. A constant supply of real-time data from building sites is made possible by IoT and sensor technology as the data contains a range of aspects, such as resource use, temperature, and humidity. As a consequence of better decision-making, risk mitigation, and process optimization by construction managers, projects are managed more effectively and efficiently. The operational lifespan of structures and infrastructure is also affected by IoT beyond the development stage. Predictive maintenance, effective energy management, and improved occupant comfort are all made possible by smart buildings using IoT sensors. A sustainable and user-centered built environment is created by using a comprehensive approach to design, construction, operation, and maintenance.

Augmented Reality (AR) and Virtual Reality (VR) in Construction

Enhancing visualization, training, and quality control via the use of AR and VR technology in the construction industry is a game-changing development. By introducing novel methods of comprehension and participation, these immersive technologies revolutionize the way building operations are performed. Stakeholders can utilize AR to overlay virtual models onto actual environments, providing a unique perspective of design intentions. This interactive visualization supports collaboration, early-stage decision-making, and design validation. Similarly, construction project planning can benefit greatly from VR training as stakeholders may explore site layouts, evaluate hazards, and adjust project timetables using VR simulations that immerse them in realistic project situations. This virtual rehearsal improves judgment, reduces mistakes, and speeds up project execution.

Industry 4.0 in Construction: Key Enablers

Key enablers that offer the required framework for the integration of cutting-edge technologies are essential for realizing Industry 4.0's disruptive potential in the construction industry. Real-time data transmission across diverse devices, sensors, and systems is made possible by high-speed internet, reliable wireless networks, and frictionless data exchange frameworks. The foundation of the digital ecosystem is this infrastructure, which enables diverse technologies to work together, share knowledge, and coordinate actions. Due to improvements in connection and network infrastructure, the modern building site is turning into a hotspot for digital connectivity.

The ongoing interaction between various equipment, sensors, and construction management systems is made possible by high-speed internet and wireless networks, which are already common components. This flexible connectivity improves project collaboration, simplifies information flow, and speeds up decision-making. In the construction context, wireless communication technologies are essential for real-time data sharing. Wireless communication enables the immediate transfer of data from wearable devices worn by employees to sensors built into construction equipment (Awolusi et al., 2018). This promotes collaboration among stakeholders by facilitating quick insights, enabling prompt modifications, and optimizing resource allocation.

Integration and Interoperability of Digital Systems and Tools

The smooth integration and interoperability of various digital systems and technologies are essential for the successful convergence of Industry 4.0 in the construction industry. The power of interoperability is demonstrated through the fusion of BIM and Construction Management Systems. BIM-infused data allows for a smooth transition from planning to execution by connecting together the design and management stages (Shirowzhan et al., 2020). This connection improves accuracy, facilitates cooperation, and offers a comprehensive perspective of the project lifecycle. The fundamental concepts for seamless data interchange across diverse building systems are interoperability standards. These criteria guarantee that information may be communicated without integrity or understanding being lost. Construction stakeholders ensure seamless information flow, facilitating well-informed decision-making and coordinated activities, by adhering to established standards.

Adoption of Standardization and Best Practices in Industry 4.0

Adoption of standardization and best practices is crucial for the effective implementation of Industry 4.0 in the construction industry. This enabler promotes an integrated Industry 4.0 environment within the construction industry by ensuring uniformity, interoperability, and an organized approach to deploying cutting-edge technology. A foundation for implementing Industry 4.0 in the construction industry that is widely approved is provided by international standards and guidelines. By guaranteeing the technology, processes, and data formats comply with international standards, these standards improve interoperability and harmonize procedures between various projects and organizations (Porter et al., 2014). Construction stakeholders create a common language that fosters efficiency and collaboration by accepting international standards. The application of Industry 4.0 technology in the construction industry is guided by best practices. These practices highlight successful methods for adopting, integrating, and managing technology, drawing on successful case studies and experiences. Construction companies could decrease risks, streamline processes, and boost the overall effect of Industry 4.0 projects by adhering to these proven methods. Construction companies can examine their preparation for adopting Industry 4.0 by using assessment frameworks for maturity and readiness (Ramanathan & Samaranayake, 2022). These frameworks assess a number of variables, including process alignment, talent development, and technical infrastructure. Construction stakeholders can also strategically plan and prioritize their Industry 4.0 projects by being aware of their existing situation, resulting in a well-calibrated integration process.

Collaborative Culture and Change Management in Construction Organizations

A collaborative organizational culture and successful change management are essential for implementing Industry 4.0 in construction organizations. Construction companies are more equipped to survive in the Industry 4.0 environment when they have a collaborative culture. Open communication, cross-functional cooperation, and information sharing are all encouraged in this culture. The potential of the latest technologies can potentially be collectively tapped into by construction experts, who can also share knowledge and create together by encouraging an atmosphere where information can flow freely. For Industry 4.0 technologies to be adopted smoothly and successfully, change management practices are essential (de Sousa Jabbour et al., 2018). These techniques make sure that staff members are prepared to accept change, acknowledging that the incorporation of new technology needs organizational changes. Construction organizations may minimize opposition and maximize industry 4.0 advantages by communicating clearly, integrating stakeholders, and resolving concerns.

Training Programs for Building Digital Skills in Construction

The importance of training and education programs in promoting digital skills within the construction industry cannot be emphasized in the context of Industry 4.0's shifting landscape. These programs serve as crucial catalysts in empowering the workforce to successfully manage the intricacies of cutting-edge technologies. A workforce with the skills to fully utilize Industry 4.0 technologies, such as AI, data analytics, and robots, is required due to their fast integration (Abbasi et al., 2022). By providing specialized learning experiences that improve the digital savvy of construction professionals, training and education programs close this gap. These programs provide a wide range of educational options, including conventional classroom instruction, online courses, and practical workshops. In addition to educating participants about the principles of digital technology, they also explore real-world applications in the context of building operations. Participants acquire hands-on experience through interactive seminars and realistic simulations, enabling them to use digital tools and systems successfully. Additionally, Industry 4.0's collaborative nature calls for transdisciplinary expertise. By bringing together experts from different fields, such as architects, engineers, project managers, and data analysts, training and education programs stimulate cross-functional cooperation. These professionals then work together to research and implement digital advances in building projects.

References

Abanda, F.H. and Byers, L. (2016). An investigation of the impact of building orientation on energy consumption in a domestic building using emerging BIM (Building Information Modelling). *Energy*, 97, 517–527. https://doi.org/https://doi.org/10.1016/j.energy.2015.12.135.

Abbasi, M.F., Bilal, M. and Rasheed, K. (2022). Role of Human Intuition in AI Aided Managerial Decision Making: A Review. *2022 International Conference on Decision Aid Sciences and Applications, DASA 2022*. https://doi.org/10.1109/DASA54658.2022.9765153.

Abdul Hannan Qureshi, Wesam Salah Alaloul, Manzoor, B., Musarat, M.A., Saad, S. and Ammad, S. (2020). Implications of Machine Learning Integrated Technologies for Construction Progress Detection Under Industry 4.0 (IR4.0). *2020 Second International Sustainability and Resilience Conference: Technology and Innovation in Building Designs, 0*. https://doi.org/10.1109/IEEECONF51154.2020.9319974.

Akter, S., McCarthy, G., Sajib, S., Michael, K., Dwivedi, Y.K., D'Ambra, J. and Shen, K.N. (2021). Algorithmic bias in data-driven innovation in the age of AI. *International Journal of Information Management*, 60, 102387. https://doi.org/https://doi.org/10.1016/j.ijinfomgt.2021.102387.

Ammad, S., Alaloul, W.S. and Saad, S. (2022). Challenges of cyber-physical systems implementation in construction industry. *In*: D.W.S. Alaloul (Ed.), *Cyber-Physical Systems in the Construction Sector* (1st Edn.), p. 19. CRC Press. https://doi.org/10.1201/9781003190134-10.

Awolusi, I., Marks, E. and Hallowell, M. (2018). Wearable technology for personalized construction safety monitoring and trending: Review of applicable devices. *Automation in Construction*, 85, 96–106. https://doi.org/https://doi.org/10.1016/j.autcon.2017.10.010.

Cipresso, P., Giglioli, I.A.C., Raya, M.A. and Riva, G. (2018). The past, present, and future of virtual and augmented reality research: A network and cluster analysis of the literature. *Frontiers in Psychology*, 9(NOV), 1–20. https://doi.org/10.3389/fpsyg.2018.02086.

de Sousa Jabbour, A.B.L., Jabbour, C.J.C., Foropon, C. and Godinho Filho, M. (2018). When titans meet – Can industry 4.0 revolutionize the environmentally sustainable manufacturing wave? The role of critical success factors. *Technological Forecasting and Social Change*, 132, 18–25. https://doi.org/https://doi.org/10.1016/j.techfore.2018.01.017.

Dixit, U.S., Hazarika, M. and Davim, J.P. (2017). Emergence of production and industrial engineering. *In:* U.S. Dixit, M. Hazarika and J.P. Davim (Eds.), *A Brief History of Mechanical Engineering*, pp. 127–146. Springer International Publishing. https://doi.org/10.1007/978-3-319-42916-8_6.

Duggineni, S. (2023). *Impact of Controls on Data Integrity and Information Systems*. July. https://doi.org/10.5923/j.scit.20231302.04.

Huang, B., Wang, X., Kua, H., Geng, Y., Bleischwitz, R. and Ren, J. (2018). Construction and demolition waste management in China through the 3R principle. *Resources, Conservation and Recycling*, 129, 36–44. https://doi.org/https://doi.org/10.1016/j.resconrec.2017.09.029.

Jiang, Z., Yuan, S., Ma, J. and Wang, Q. (2022). The evolution of production scheduling from Industry 3.0 through Industry 4.0. *International Journal of Production Research*, 60(11), 3534–3554. https://doi.org/10.1080/00207543.2021.1925772.

Meier, S., Klarmann, S., Thielen, N., Pfefferer, C., Kuhn, M. and Franke, J. (2023). A process model for systematically setting up the data basis for data-driven projects in manufacturing. *Journal of Manufacturing Systems*, 71, 1–19. https://doi.org/https://doi.org/10.1016/j.jmsy.2023.08.024.

Menezes, B.C., Kelly, J.D. and Leal, A.G. (2019). Identification and design of Industry 4.0 opportunities in manufacturing: Examples from mature industries to laboratory level systems. *IFAC-PapersOnLine*, 52(13), 2494–2500. https://doi.org/https://doi.org/10.1016/j.ifacol.2019.11.581.

Naeem, U.H., Bilal, M., Syed, F., Rasheed, K. and Saad, S. (2022). A Multilayer Encryption Model to Protect Healthcare Data in Cloud Environment. *2022 International Conference on Data Analytics for Business and Industry, ICDABI 2022*. https://doi.org/10.1109/ICDABI56818.2022.10041708.

Pan, Y. and Zhang, L. (2023). Integrating BIM and AI for smart construction management: Current status and future directions. *Archives of Computational Methods in Engineering*, 30(2), 1081–1110. https://doi.org/10.1007/s11831-022-09830-8.

Porter, C.H., Villalobos, C., Holzworth, D., Nelson, R., White, J.W., Athanasiadis, I.N., Janssen, S., Ripoche, D., Cufi, J., Raes, D., Zhang, M., Knapen, R., Sahajpal, R., Boote, K. and Jones, J.W. (2014). Harmonization and translation of crop modeling data to ensure interoperability. *Environmental Modelling & Software*, 62, 495–508. https://doi.org/https://doi.org/10.1016/j.envsoft.2014.09.004.

Qureshi, A.H., Alaloul, W.S., Manzoor, B., Musarat, M.A., Saad, S. and Ammad, S. (2020). Implications of machine learning integrated technologies for construction progress detection under industry 4.0 (IR 4.0). *2020 Second International Sustainability and Resilience Conference: Technology and Innovation in Building Designs (51154)*, pp. 1–6. https://doi.org/10.1109/IEEECONF51154.2020.9319974.

Ramanathan, K. and Samaranayake, P. (2022). Assessing Industry 4.0 readiness in manufacturing: A self-diagnostic framework and an illustrative case study. *Journal of Manufacturing Technology Management*, 33(3), 468–488. https://doi.org/10.1108/JMTM-09-2021-0339.

Rasheed, K., Saad, S., Shahzad, L., Ammad, S., Ali, A. and Badshah, I. (2021). Application of Gesture Data Recognition in a Human-Interactive Leap Motion Sensor Chair. *2021 International Conference on Data Analytics for Business and Industry, ICDABI 2021*. https://doi.org/10.1109/ICDABI53623.2021.9655819.

Rasheed, K., Shahzad, L., Saad, S., Khan, H.A., Ahmed, W. and Sadiq, T. (2021). Parking Guidance System Using Wireless Sensor Networks. *2021 International Conference on Decision Aid Sciences and Application, DASA 2021*. https://doi.org/10.1109/DASA53625.2021.9682213.

Rasheed, K., Ammad, S., Said, A.Y., Balbehaith, M., Oad, V.K. and Khan, A. (2022). Application of Smart IoT Technology in Project Management Scenarios. *2022 International Conference on Data Analytics for Business and Industry, ICDABI 2022*. https://doi.org/10.1109/ICDABI56818.2022.10041674.

Roblek, V., Meško, M. and Krapež, A. (2016). A complex view of Industry 4.0. *SAGE Open*, 6(2). https://doi.org/10.1177/2158244016653987.

Saad, S., Alaloul, W.S., Ammad, S., Qureshi, A.H., Altaf, M. and Rasheed, K. (2020). Design Phase Carbon Emission Prediction Using a Visual Programming Technique. *2020 2nd International Sustainability and Resilience Conference: Technology and Innovation in Building Designs*. https://doi.org/10.1109/IEEECONF51154.2020.9319978.

Saad, S., Alaloul, W.S., Ammad, S., Altaf, M. and Qureshi, A.H. (2021). Identification of critical success factors for the adoption of Industrialized Building System (IBS) in Malaysian construction industry. *Ain Shams Engineering Journal*, 13(2), 101547. https://doi.org/https://doi.org/10.1016/j.asej.2021.06.031.

Saad, S., Alaloul, W.S. and Ammad, S. (2022). Role of cyber-physical systems in smart cities. *In*: D.W.S. Alaloul (Ed.), *Cyber-Physical Systems in the Construction Sector* (1st Edn.), p. 19. CRC Press. https://doi.org/9781003190134.

Saad, S., Alaloul, W.S., Rasheed, K. and Ammad, S. (2022). Modeling and simulation of construction cyber-physical systems. *In*: *Cyber-Physical Systems in the Construction Sector*, pp. 88–110. CRC Press.

Sachs, J.D. (2019). The ages of globalization. *In*: *Geography, Technology, and Institutions*, pp. 129–168. Columbia University Press. https://doi.org/doi:10.7312/sach19374-008.

Salem, S.S. (2023). A mini review on green nanotechnology and its development in biological effects. *Archives of Microbiology*, 205(4), 128. https://doi.org/10.1007/s00203-023-03467-2.

Shirowzhan, S., Sepasgozar, S.M.E., Edwards, D.J., Li, H. and Wang, C. (2020). BIM compatibility and its differentiation with interoperability challenges as an innovation factor. *Automation in Construction*, 112, 103086. https://doi.org/https://doi.org/10.1016/j.autcon.2020.103086.

Stavropoulos, P., Foteinopoulos, P., Papacharalampopoulos, A. and Bikas, H. (2018). Addressing the challenges for the industrial application of additive manufacturing: Towards a hybrid solution. *International Journal of Lightweight Materials and Manufacture*, 1(3), 157–168. https://doi.org/https://doi.org/10.1016/j.ijlmm.2018.07.002.

Teng, S.Y., Touš, M., Leong, W.D., How, B.S., Lam, H.L. and Máša, V. (2021). Recent advances on industrial data-driven energy savings: Digital twins and infrastructures. *Renewable and Sustainable Energy Reviews*, 135, 110208. https://doi.org/https://doi.org/10.1016/j.rser.2020.110208.

Wu, Z., Luo, L., Li, H., Wang, Y., Bi, G. and Antwi-Afari, M.F. (2021). An analysis on promoting prefabrication implementation in construction industry towards sustainability. *International Journal of Environmental Research and Public Health*, 18(21). https://doi.org/10.3390/ijerph182111493.

CHAPTER

5

Automation in Construction Industry

Ahmad Zaland, Syed Saad, Kumeel Rasheed and Syed Ammad

Introduction

The scope of automation in construction is vast, covering all phases of construction from initial planning and design, through construction of the installation, its use and maintenance, to the ultimate demolition and recycling of buildings and technical structures. New technologies have been developed in the construction business as a result of recent advancements in the disciplines of robotics and computer intelligence. Numerous innovative technologies and ministries have been created in Japan, a world leader in robotics and automation, which have aided the construction sector in reducing fatalities, construction costs, and design times while increasing productivity (Tajenkar, 2017).

A construction point is largely dynamic due to rapid-fire changes and numerous moving corridors. Common problems on-point are poor communication between design phases and actors, lack of information sharing, misgivings, dependable deliveries, and material overflows. The construction industry has also plodded with high material destruction, costs, flat productivity, safety and deficit labor (Ammad, Alaloul, et al., 2021).

Construction systems are distinctive in a variety of aspects, including various figure constraints, job activities, and design costs. This indicates that it will be difficult to implement robotization in the construction sector. Since the landscape of the construction business is always changing and many jobs are conducted in the same way every time, implementing robotization is a significant difficulty. Even if research into building automation started decades ago, one could question why this industry doesn't have a high level of robotization (Tajenkar, 2017).

Manufacturing companies are behind other efforts in digitization and robotics. Digitization is key to robotics. According to Richards (1994), the development of computer models containing all the necessary information to create a database for use in the following production steps in manufacturing requires the presence of robots to a large extent. Garcia de Soto says the reasons for the decline in productivity in the construction industry are many, such as resistance to change stemming from the fact that the construction industry is indeed traditional and it suffers from poor performance, data, low exchange rates, and high growth conditions. Wu et al. (2016) also mentioned that construction industry is seen as a technology-free industry. Reasons for slow discovery may include longevity, high product diversity and complexity, multiple boundaries, manufacturing fixedness, reluctance to use alternative methods, and low budgets for insight and development. The global construction industry has not experienced much technological advancement.

An Overview of Construction Automation

It's easy to imagine robots and automated tools flying around the manufacturing space as part of a distant learning future, but the truth is that the important ways of using these tools have been developed has itself been glorified, and presented the concept of automation for centuries. Examples of prefabricated architecture differed by more than 2000-fold from fragments, ranging from prefabricated methods used to build clay armies in the 3rd century B.C.

However, once the first artificial robots were built in the 1950s and put to use by the automobile sector in the 1960s, ultramodern construction robotization with robotics took off. Construction robotics emerged in the 1960s and 1970s as plant robotization moved throughout the artificial world. Japan developed construction robotization and robotics in the 1970s and 1980s in response to a shortage of labor in the construction industry brought on by a rising population and uninterested young employees. Robots and remote-controlled machines were developed by Japanese armature and engineering firms like the Shimizu Corporation, Obayashi Corporation, and Takenaka Corporation for digging, handling tools, pouring and finishing concrete, fireproofing, building walls, installing rebar, and other construction-related tasks.

Types of Automation in Construction

Automation in the construction industry focuses on the implementation of technology and ministry to jobs that have historically been done by people. The construction sector uses a variety of automation techniques.

Off-site Construction Automation

Off-site construction automation describes practices that transform the construction process into a more modernized, automated manufacturing environment. Off-site construction includes several related yet distinct terms such as prefabrication, volumetric and panelized modular construction, and precast. Construction processes are moving away from job sites to manufactories within a managed setting that prioritizes automation, artificial intelligence (AI) integration, digital product workflows optimization, and design for manufacture and assembly (DfMA) strategies.

In the construction business, off-site automation is more prevalent than automated off-site processes, and the proximity to manufacturing has made it easier to transfer direct manufacturing technologies, with one important exception. Automated production lines are typically utilized in manufacturing for high-volume products when the part size, shape, and assembly order are consistent throughout tens of thousands of units. While assemblies of cultivated corridor are used in the construction of buildings, roads, and islands, the variety of tools and processes, and the necessary variation between elements and systems, present a special challenge for tooling (the configuration of automated equipment in a product line). The product line must be automated while being flexible enough to adapt to variation.

Plant automation requires a significant investment, but over time it can save money, time, and resources while improving control of quality as well as quality confirmation and providing safer, more pleasant environments for employees by eliminating many repetitive tasks common to standard construction processes. Plant-based building may reduce waste, use less water, reduce functional energy usage and dust pollution, and maximize material utilization, physical activity, and recycling. Additionally, it will significantly contribute to addressing the worldwide need for structures and architecture when paired with automated procedures. A number of the largest and most audacious automated construction facilities are designed to operate mostly without human oversight.

The famous quote by the American professor and pioneer in the study of leadership Warren Bennis, "The crop of future generations will employ just two employees, a man and a dog," comes to mind. In order to feed the dog, the guy will be present. To prevent the man from touching the costume, the dog will be present (Davis, 2022).

On-site Construction Automation

With a few exceptions, plant-based automation in construction may be seen as an innovation transfer from manufacturing, where automated tooling is set up to create building blocks rather than finished goods. On-site construction automation continues to bring a variety of distinct opportunities. A rich field for research, new commercial endeavors, and start-ups is the development and planting of infrastructure, which requires new tools and procedures. Construction automation equipment designed for on-site use must be portable enough to reach project sites, and be able to be set up, operated, and taken down in order to proceed to the next job. With a view towards a fully or semi-automated future, some existing equipment, comparable as heavy earth-moving machines, has been converted and new equipment is being created less frequently. Devices used in on-site construction and off-site construction as mentioned in Table 5.1.

Table 5.1: Devices used in on-site construction and off-site construction

S. No.	On-site construction devices	Off-site construction devices
1.	Mobile Robotics Platform On the building site, these gadgets may move on their own and carry out operations including site scanning, data transmission, and layout.	Automated Manufacturing Equipment Specialized equipment used in off-site production facilities to create prefabricated pieces or modules, like as walls, floors, or volumetric units.
2.	Construction Site Scanning and Layout Robots The purpose of these robots is to scan the building site and aid in the accurate placing and arrangement of items.	Robotic Assembly Arms Robotic arms with automation are used in industrial facilities to precisely assemble and weld components.
3.	Heavy Earth-Moving Machinery Heavy earth-moving equipment is used for grading, transporting materials, and excavating on construction sites. Examples include bulldozers, loaders, and excavators.	Computer Numerical Control (CNC) Machines Various structural pieces are cut, shaped, and drilled using CNC machines to streamline the manufacturing process.
4.	Portable Automated Outfit Specific equipment created for on-site automation, like as robotic arms, automated concrete placing systems, or rebar placement machines.	Digital Design and Modeling Software Advanced design and modeling tools for off-site prefabricated components.
5.	Automatic Vehicles Autonomous cars may be used at some construction sites for logistics and material transportation.	Transportation Equipment Transporting prefabricated components from the production plant to the building site using trucks, trailers, or other types of vehicles.
6.	Augmented Reality (AR) Devices For on-site visualization, quality control, and delivering building instructions, AR glasses or headsets may be employed.	Material Handling Machinery Equipment for effectively handling and moving goods inside the off-site manufacturing plant.
7.	Handled Digital Devices To monitor progress, organize activities, and access digital building blueprints, use tablets or smartphones with construction management apps.	Quality Control Devices Sensors and cameras for inspection are among the equipment and instruments used to check the quality of prefabricated components.

Early examples of on-site building robotization frequently compromised the unity of the structure and reverberated in erecting methods designed specifically to deal with them. Currently, there is a different approach to automated building that uses standardized fundamentals and supports changes among units. An automated system that pours concrete underpinning, for instance, minimizes repetitive chores on the jobsite, permits performance-driven flexibility in the placement of rebar without adding unnecessary expense, and reduces waste through putting stuff exactly where it is needed.

Boston-based launch vehicle on-site construction automation systems are created by NeXtera Robotics; one such system is the Oliver construction site scanning and drawing robot. Similar to the plasterboard installation robots that NeXtera is creating, some on-site construction automation tools may also be suited to on-site prefabrication, although using this equipment locally can reduce the builder's shipping expenses.

Other businesses are likewise focusing on the difficult task of layout, a laborious task where precision is expected. For instance, Dusty Robotics uses mobile robotic platforms to deploy construction data gathered from a computerized model and move that information to the construction site, essentially posting the building process guidelines right away on the bottom of the building itself and reducing time and labor costs while enhancing delicacy (Davis, 2022).

Advantages of Automation in Construction

Automation is assisting the building industry's fast evolution by making operations more effective, secure, and cost-efficient. Intelligent document processing (IDP) is one such automation technology that is advancing the digital revolution of the sector.

This cloud-native solution streamlines the whole document lifecycle, from data gathering and pre-processing through post-processing and reporting, by utilizing cutting-edge AI and machine learning (ML) technology. The IDP platform from Rossum not only makes document processing easier, but it also substantially reduces the amount of human work required and offers many additional advantages for your projects.

In a highly competitive industry, having the appropriate automation for quicker and more accurate document processing is essential. Learn about the eight important advantages of IDP construction automation, including how it lowers mistakes and raises client satisfaction (Solomom, 2023).

Enhanced Effectiveness

Contracts, applications, and permit operations may all be streamlined for construction companies by using automation. Teams may check data more simply and expedite transactions when using IDP for document automation. This improves your team's general accuracy and efficiency while handling papers.

Lower Expenses

Companies may now process documents more quickly thanks to technology, which lowers labor costs and time-consuming delays. Businesses may reduce waste and save money by using IDP to replace labor-intensive and error-prone procedures like manual data input.

Additionally, staff may concentrate more on activities that generate revenue by spending less time on tedious paperwork responsibilities.

More Accurate Rates

Automation lowers the chance of human mistake, which raises accuracy rates. IDP combines cutting-edge ML and AI algorithms to correctly extrapolate data from documents of any design, making it the most exact and adaptable data capture solution. This prevents costly future rework and avoids human mistakes like entering wrong information.

Increased Safety

Throughout all phases of a project's lifetime, protecting sensitive data is essential for safe teamwork between teams, contractors, and suppliers (Bilal et al., 2022). Automated solutions with strong security and privacy protections improve data integrity and guard against loss, destruction, manipulation, and unauthorized access.

Better Coordination

In construction projects, documentation accuracy is crucial; even little errors in the workflow may be quite expensive. At every step of the project lifecycle, all documents will be correct, arranged, and simple to discover thanks to automated document processing.

Contracts, meeting minutes, and reports may all be automated to ensure that projects are always up-to-date and that stakeholders have easy access to the information they want.

Simpler Reporting and Monitoring

The capacity to track each document's journey till it reaches its goal, be it an ERP, spend administration system, or DMS, is a big benefit of utilizing IDP for document automation.

Additionally, you may leverage the real-time analytics provided by IDP's built-in reporting and dashboards to comprehend your document processing activity better.

The automation has developed into a crucial tool for companies looking to reduce costs and boost production (Sánchez & Hartlieb, 2020). This is accomplished by intelligent document processing, which streamlines papers in a quarter of the time it usually does. Along with time and money savings, your organization's accuracy and security also improve.

It's not as hard as it would appear to create an automated solution that is accurate, adaptable, and scalable. The task can be completed quickly and effortlessly with the correct partner and equipment.

Effortless Cooperation

Because automation makes it easier to share papers and access information while requiring less manual follow-up, it may significantly increase cooperation. This gives teams more time to concentrate on their tasks rather than waste it looking for information.

Collaboration between departments is improved by using a centralized document platform to make it straightforward to distribute files among team members or between organizational divisions.

Case Studies for Automation in Construction

Some of the case studies that show how automation is being used in the construction sector are following.

Automated Masonry with SAM (Semi-Automated Mason)

Sam100 is a robot for on-site masonry building that lays bricks semi-automatically. Sam100 was developed to help with the taxing work of lifting and putting bricks, boosting the output of masons by 3–5 times that of which they could achieve before while lowering the lifting by 80%.

Construction workers were not intended to be replaced by Sam100; rather, it was created to improve their skills and reduce the hazards to their health and safety. The majority of the site preparation and final wall quality inspection will still be done by the mason.

At the World of Concrete trade event in 2015, when it was first introduced, this semi-automated mason robot earned the Most Innovative Product Industry Choice Award. In 2017, the Sam100 made a second appearance at the World of Concrete, however the new 2.0 version is substantially quicker than the older one (Jones, 2019).

Benefits of SAM

For on-site masonry work, SAM is the first bricklaying robot. SAM collaborates with the masons, greatly enhancing the amount of work that a single mason can accomplish. The following advantages of the SAM system are offered.

- Lower installation costs – labor savings of up to 50%
- Increase output by 3–5 times
- Cut back on lifting by 80%
- Designed to assist the mason in their task
- Constant advancements using manufacturing data
- Fewer safety and health issues
- Job cuts of more than 20%
- Less physical stress on the crew and mason since the mason uses the system to tool joints, operate the system, and ensure good wall quality

- The completed wall product's is dependably of high quality
- Reduced effects of health and safety on the workforce
- Consistency in performance and output rate
- Predictable task estimation.

Sam100 is capable of placing 3000 bricks in 24 hours as opposed to the typical construction worker's 500 bricks per day. Many employees are worried that this may jeopardize their ability to keep their jobs, but Sam100 wasn't developed to displace people from their jobs. Having Sam100 on a jobsite is like to adding another mason to your crew, said Andy Sneed, CEO of Wasco. Sam100 concentrates on the heavy lifting, placing brick after brick, freeing the mason employees to concentrate more on quality assurance and other aspects of the wall.

According to operations manager, Zak Podkaminer, "We don't anticipate building sites becoming entirely automated for decades, if not centuries". This is about partnership between human employees and robots. SAM picks up the bricks, coats them with mortar, and then fastens them to the wall. Still, it needs a mason working beside it. SAM is just there to carry out the difficult tasks (Jones, 2019).

Robotic Rebar Tying

A rebar tying robot created by US-based entrepreneur Stephen Muck can shorten the amount of time spent on a single building site. Tying rebar is one of the most laborious, boring, and back-breaking tasks on a construction site. A rebar-tying robot, however, may be a time-saving substitute, according to a US entrepreneur.

Stephen Muck, the creator and CEO of Advanced Construction Robotics in Pennsylvania, invented the TyBot. "With just one person needed to monitor, it can attach rebar at the rate of a team of six to eight construction workers," the author claims (Sweet, 2018).

The robot was used in a project in western Pennsylvania in late 2017 after being tested on bridge-building jobs. "We put the bottom layer of steel on a bridge's deck, respectively, and because it went so well, we were asked to return the following week to tie the top mat," adds Muck. Depending on how broad the bridge deck is, the motorized frame of the TyBot may stretch to a width of up to 42 meters.

The reinforcement bars are fastened together by a robotic arm as it glides around the frame, hovering above each rebar intersection. Repeating this technique, the frame slides across the bridge. To lay nearly 10,000 sq-m of reinforcement and connect about two million junctions, a group of eight to ten employees for Brayman needed roughly 7400 person-hours over the course of six months to construct the Hulton Bridge in Oakmont, Pennsylvania.

"The TyBot expedites the job while requiring fewer workers, according to Muck. The robot also improves security and health since it prevents injuries brought on by workers walking between the rebar and stooping to knot the crossings."

Another time-saving feature of the TyBot is that it could be used after hours or while construction personnel are preoccupied with other chores. This is a case of a commercial issue in the construction sector being solved by robotics (Muck, 2018).

Within an hour of arrival, it begins tying up to 1100 rebar intersections each hour on the project. It operates day or night, in good weather or bad, greatly increasing employee productivity, lowering the risk to schedules, and increasing the capacity to do more work while keeping personnel safe.

- Autonomous working area navigation without the need for pre-mapping or calibration
- Finds and binds rebar intersections
- Does not interfere with infrastructure or other construction tools
- Leased option with a knowledgeable and devoted quality control technician on staff.

Prefab Construction with Factory-based Automation

The word "construction automation" refers to the procedures, apparatus, and machinery used in the construction of infrastructure and structures. Tools are frequently employed to automate manual operations; nevertheless, in other cases, automated tooling enables the transfer of knowledge or creation of new processes specifically for construction. Building initiatives can be automated at different stages, beginning through the center around the software design phase, proceeding on to the programmed off-site as well as on-site construction, and finishing with the sharing of gathered data on the infrastructure and energy consumption of finished structures all recorded in cloud-based living models (Altaf et al., 2020). It takes a variety of key software and hardware development approaches to realize this integrated feedback loop. For instance, new robots and automated machinery, industrialized building techniques, actual time directly sensing, suggestions, and adaptability are some of the innovations and practices that are combining to make building automation a frequent reality (Davis, 2022).

Brief History of Construction Automation

It's not a novel concept to compare building procedures to those in manufacturing. There are accounts of house kits being created and distributed across the US mainland in the 1800s. For instance, in the early 1900s, businesses like Sears Roebuck created prefabricated homes that were offered through catalogues. Customers chose the design they desired, and Sears Roebuck shipped (Debs et al., 2021).

Types of Construction Automation

Off-site Construction Automation

Off-site construction automation describes techniques that bring the construction process closer to modern automated production. Prefabrication, volumetric and panelized construction with modules, and precast are a few terms that are similar but not identical when referring to off-site building. Construction operations are moved from project sites to facilities where they may be enhanced using automation, robots for industrial use, digital production processes, and development for production and assemblage (DfMA) methods. With one significant exception, the building sector employs off-site automation more often than automated on-site activities, and the close proximity to production has made it simpler to transfer direct technology to manufacturing (M. Wang et al., 2020). In manufacturing, automated production lines are frequently used for high-volume production where the product's size, shape, and assembly sequence remain similar all through thousands of units. Although it requires a substantial investment, industrial automation could eventually save resources, time, and money. It can also enhance control and quality assurance and increase worker comfort and safety by eliminating many of the frequently repetitive activities seen in conventional building processes (Ammad et al., 2020). By minimizing waste production, water usage, operating energy use, and dust pollution, as well as by maximizing material use, reuse, and recycling, factory-based building may have a positive impact on the environment. Additionally, it will significantly contribute to addressing the need for infrastructure and structures on a worldwide scale when paired with automated procedures. The most cutting-edge automated factories for construction are made to run with little to no human involvement (Davis, 2022).

On-site Construction Automation

With certain notable exceptions, factory-based robotics in the construction industry may be considered as a technological transition from production, where mechanized tooling is set up to produce building parts rather than finished commodities. On-site construction technology, however, presents unique, brand-new opportunities and challenges. The creation and implementation of technology become less of a straight transfer and need new equipment and procedures, creating a fertile area for research, new business endeavors, and start-ups. Construction automation tools created for use at the job site must be transportable enough to be put up, used, and then dismantled to go on to the next project (Saad et al., 2022). Some equipment has been modernized, such as massive earthmoving machinery, while fresh machinery is being built with a fully or a semi-automated destiny in mind.

Early instances of on-site automation prompted the development of particular building systems to work with them, frequently reducing the individuality of the structure. There is currently an additional effort at automated manufacturing that makes use of standardized parts and accommodates variations across units. Automated machinery that installs reinforcement into concrete, for example, reduces repetitive tasks on the jobsite, offers driven-by-performance versatility in reinforcement placement without incurring additional expenditures, and reduces waste by positioning material exactly where it is required.

A company based in Boston, The Oliver construction site-scanning and layout robot is one of the on-site construction automation systems produced by NeXtera Robotics. Some on-site construction automation tools, like the plasterboard installation robots NeXtera is creating, may also be used in off-site prefabrication, but using them on-site can save the builder money on shipping (Davis 2022).

Overview of Prefabricated Building

Prefabricated buildings, often known as prefabs, are ones that include components (such as walls, roofs, and bottoms) that are produced in an assembly line or manufacturing factory. These components may be partially or entirely built in a factory, and brought to the spot. This method of building is chosen since it is affordable, quick to reverse, and reusable. Prefabricated buildings are frequently used for temporary construction sites, offices, medical camps, evacuation centers, seminaries, apartment buildings, and single-family homes (Mohsen Alawag et al., 2023).

Prefabrication is more efficient than traditional on-site building because it allows for more precise manufacture via a product line (Jiang et al., 2023). Most constructions have repeated roofs, walls, and bottom portions, therefore by connecting a series of processes together, a method of production may be created. The manufacturing process may be improved by studying and improving these processes.

Many prefabricated buildings that have been constructed throughout history may be traced back to wandering periods in which people were emigrating to explore new regions. With colonization came the necessity for portable camps and homes. However, the requirement for prefabricated buildings was less pressing in the past because there was no requirement for further growth after the colony was settled.

The twentieth century was the first time the system underwent noteworthy development. Due to the significant need for men in the production of war items, there was less labor available for building during the first and second world wars. Casing shortages resulted from this and persisted into the postwar era. In order to fulfill the demand, necessary casing construction types were taken into consideration.

Prefabricated structures must adhere to the same structural rules as infinite constructions even though they are made from modules. Because they differ from a single nation or county to another, structural canons are difficult to adhere to. The directors of infinite and prefabricated structures struggle with this element of building. The world's International Building Code, which differs between nations but is regularly updated to reflect new advancements, must be followed for modular or prefabricated constructions to be permitted in the United States.

The building site is examined to determine the kind of soil required for the groundwork of a precast structures before it is installed. The point is checked by an outside party to make sure it complies with the genuine nation, and global structure canons in order to receive further blessing. Modular firms must get structure permits, distance permits, and residence permits after passing the colorful tests.

Prefabricated structures are continually being built in place of conventional building as a result of advancements in and standardization of construction and structural canons, as well as the increased need for housing and office space (Chen et al., 2023). By digitally expressing the structural properties, modeling tools and methods akin to BIM (Building Information Modelling) assist engineers, architects, and contractors. BIM makes it possible for the assembly line to run efficiently, minimizing the risks for businesses involved in prefabricated building (Ammad, Saad, et al., 2021).

Benefits of Using Prefabricated Building

Manufacturers, builders, and end users all gain greatly from prefabricated structures. If the planners are proficient in project management, the notion of splitting up tasks on- and off-site allows for more versatility in the timeline of the project and expenses. The benefits of optimized assembly lines are also present in off-site fabrication. Prefabricated structures will benefit from market potential as they follow the green sustainability Trendmarket share in the industry, housing and non-residential, is predicted to rise in the future years.

Rapid Off-site Construction

Among the main benefits of using a production line approach is the fast reversal. Separate tasks and defined functional sequences help construction workers perform more efficiently than conventional methods. Some operations can also be automated. With regard to design planning, prefabricated construction can be implemented quickly because some of the conditioning can be done concurrently. As an illustration, take point clearing and foundation construction. Walls, roofs, floors, and finishing must wait until the foundation is properly cured before being built. With the help of prefabrication styles, a building can have up to 90% finished states upon delivery. A fully installed system can be delivered within days or weeks after initial contact.

Resistance to Natural or Uncontrollable Factors

Construction work faces challenges from weather conditions, which are out of human control and manipulation powers. Slowing down systems, leading to delayed completion dates and disorganized workflows. A key element to consider in a design, companies generally look at the weather first.

Unlike construction systems that are never ending, rainfall has little impact on the assembly of prefabricated structures since 90% of it is done in controlled conditions. Prefabricated structures, put together in a controlled factory setting, enable the construction of new installations during rainfall conditions that would hinder conventional building methods.

The production of prefabricated buildings helps curb these external components. Structural factors can be made under further controlled conditions, uninfluenced by external terrain. Some of the styles that are minimally affected by unbridled factors include protect welding and precast concrete manufacturing (Khedmatgozar Dolati & Mehrabi, 2021) .

Higher Quality and Consistency

Since structural components are created with repeated characteristics, quality control is easier to execute than on-site building. Dimensions and tolerances of components may more easily be standardized. For a typical construction component that yields fixed dimensions, the molds, formworks, and temporary fasteners are the same. Additionally, if two prefabricated structures are erected in separate areas and use the same components, their quality is more likely to be constant. This is so that site circumstances closer to the construction are less of an issue.

Time and Efficiency

Prefabricated structures cost as much as or more than conventional construction. The main benefits of using it are time saving and increased effectiveness. Traditional building projects can often be wrapped up in six to nine months. Prefabrication manufacturers can finish these projects in half the time, with the same quality, leading to cost savings and rapid structural implementation (Saad et al., 2020). Prefabricated structures have the benefit of having components and features that are customized and unique to each project, necessitating precision engineering and design.

Breaking Down and Reusability

Some prefabricated structures are intended to be transient. These are preferable when project-based work is involved, such as when building work, remote healthcare services, research, etc. Prefabricated structures are simple to deconstruct and move to other locations. Additionally, this capability ensures that the job site is maintained and altered as little as possible.

Environmentally Friendly

This advantage results from the method's effective use of raw materials and reusability. Conventional construction produces more waste materials and transient elements that are discarded after construction, including formwork, transient fasteners, jigs, and fixtures (Mohsen Alawag et al., 2023). On-site building typically results in permanent structures. The building will remain vacant after its intended usage until it is either converted or torn down. Because of their mobility, prefabricated, modular structures are simpler to reuse (Burry et al., 2020).

Methodology

The PERI AutoPilot system automates the installation and disassembly of concrete formwork through the use of cutting-edge robotic technology. To accomplish its goals, the system depends on real-time data collecting, powerful sensors, and accurate location algorithms. The process involves the critical stages listed below:

- *Data Collection:* The system compiles crucial construction information, such as blueprints, design specifications, and necessary formwork specifications.
- *Robotic Setup:* The powerful AI and computer vision capabilities of the robotic arms are designed to recognize the precise formwork components needed for each building segment.
- *Formwork Positioning:* In accordance with the construction schedule and design specifications, the robotic arms precisely place and modify the formwork components.
- *Pouring of Concrete:* After the formwork is set up, the system makes sure that it is precisely aligned and stable before the concrete is poured.
- *Curing and Dismantling:* The PERI AutoPilot system controls the curing process once the concrete has hardened. The technique prepares the formwork for the following building phase by disassembling and removing it once the material has dried.
- *Features:* The PERI AutoPilot system provides a number of vital features that enhance the formwork process.
- *Automated Formwork Setup:* By accurately positioning and adjusting the formwork pieces autonomously, the robotic arms lessen the requirement for manual labor.
- *Real-time Data Feedback:* The technology provides real-time data on the placement of the molds and the concrete pour, allowing quick adjustments and minimizing errors.
- *Concrete Pouring Accuracy:* The robotic arms assure concrete placing accuracy, lowering the possibility of errors and boosting structural integrity.
- *Intelligent Formwork Removal:* The method effectively removes the formwork once the concrete has dried, allowing for quicker transitions between building phases.
- *Building Information Modelling (BIM):* BIM is one construction technology that PERI AutoPilot can smoothly connect into, improving coordination and communication among project stakeholders.

Benefits of Using PERI AutoPilot

Using PERI AutoPilot in construction projects has a number of benefits, including:

- *Enhanced Efficiency:* The automated formwork procedure uses less manual labor, hastening project schedules, and advancing construction.

- *Increased Accuracy:* By carefully placing formwork and pouring concrete, the robotic arms minimize the chance of mistakes and redo.
- *Enhanced Safety:* The technology eliminates the need for humans to conduct physically taxing and sometimes dangerous activities by automating the installation and removal of formwork.
- *Cost Savings:* The effectiveness of the system and the decreased need for labor result in cost savings in construction projects.
- *Quality Control:* Proactive quality control methods are made possible by real-time data feedback, guaranteeing that the construction adheres to design standards.
- *Sustainability:* The methodology helps to promote more environmentally friendly construction methods by maximizing the use of formwork and minimizing material waste.
- *Scalability:* From small-scale to large-scale developments, PERI AutoPilot may be adapted to a variety of building projects.

Automation in Construction: Planning and Scheduling

A new era of effectiveness, accuracy, and smooth project management has arrived due to automation in construction planning and scheduling. Utilizing technology and digital tools is essential to overcoming obstacles and achieving the best results as the demands on building projects become more complicated and large-scale (Ammad, Saad, et al., 2021). By utilizing the power of AI, sophisticated analytics, robotics, and real-time data management, the integration of automation in construction planning and scheduling revolutionizes conventional practices.

Effective planning has been associated with improved construction project performance, with advantages including cost and time savings, a more precise definition of the project's scope, and a decrease in the frequency of change orders. Planning was described as "before taking action, the process of selecting exactly what is needed and how to accomplish it" by Shapira Laufer and Shenhar. But frequently, building projects frequently lack proper planning due to factors like a supposedly lack of time, or a lack of organizational knowledge and motivation to complete preplanning chores among the most frequently stated. Because research has shown that ineffective planning leads to delays, cost overruns, and owner unhappiness, it is crucial to address this issue. Poor planning has detrimental effects on owners and contractors alike (Lines et al., 2015).

Construction and project managers agree that having an organized plan is the key factor in their work. Assessing resources, defining work assignments, and understanding the building sequences are all part of it. Planning a construction project leads to improved project outcomes, such as reducing cost overruns and finishing the project within the allotted time frame. BIM, which allows for a four-dimensional image of the project, is an automation in construction tool that simplifies planning (W.-C. Wang et al., 2014). According to Abanda, BIM gives the planner a scheduling image that may show the structure being constructed from start to finish in three-dimensional models. As it is simpler to see the order of activities, all parties participating in the project may arrange resources properly.

Planning and scheduling for construction formerly depended mainly on manual procedures, which were prone to human error, delays, and inefficiencies. However, the introduction of a solid foundation of linked digital solutions that simplify every aspect of the project lifecycle brought about by the advent of automation has changed these practices. Automation offers a complete set of tools that optimize resource utilization, foster collaboration, and guarantee adherence to strict project timelines, from conceptualization and design through execution and project handover.

Automation in construction enables interdisciplinary teams to work cooperatively, share data, and visualize the project in a virtual environment before construction ever begins. At the heart of automation is BIM. Its dynamic and interactive features make it easier to discover clashes, do precise quantity take-offs, and run simulations, giving stakeholders the information they need to make wise decisions throughout the planning stage. This all-encompassing strategy reduces disputes, rework, and material waste, which leads to cost savings and increased project effectiveness.

Furthermore, the use of AI in scheduling and planning building projects is revolutionary. Large datasets, past project performance, and a variety of limitations may all be analyzed by AI-driven algorithms to provide optimized plans that take job dependencies, resource availability, and external factors like weather conditions into account. This data-driven methodology gives project managers the ability to set realistic and doable deadlines, lowering the possibility of delays and cost overruns while maximizing resource allocation.

The value of automation in reducing possible hazards becomes clear as building projects increase in size and complexity (Pan & Zhang, 2021). Construction managers may foresee problems and take proactive measures to solve them by using predictive analytics. This helps them minimize possible interruptions and improves their risk management plans. Additionally, real-time monitoring and reporting guarantee that stakeholders have the most recent information on the status of the project, enabling prompt interventions and flexible decision-making.

Automation offers advantages beyond effective planning and scheduling. Through cloud-based collaboration tools, automation fosters improved communication and cooperation across varied teams dispersed across numerous places (Naeem et al., 2022). It promotes a transparent culture by giving stakeholders access to real-time information and updates, improving collaboration, and closing communication gaps.

Overall, automation in scheduling and planning for construction ushers in a time of increased production, cost efficiency, and quality control. The introduction of automation becomes not only advantageous but crucial for construction organizations hoping to remain competitive in a dynamic and ever-evolving market as building projects grow more ambitious and timetables more demanding. For construction companies aiming to create a future of success, sustainability, and innovation, embracing automation is more than simply a choice—it is a strategic requirement as shown in Figure 5.1.

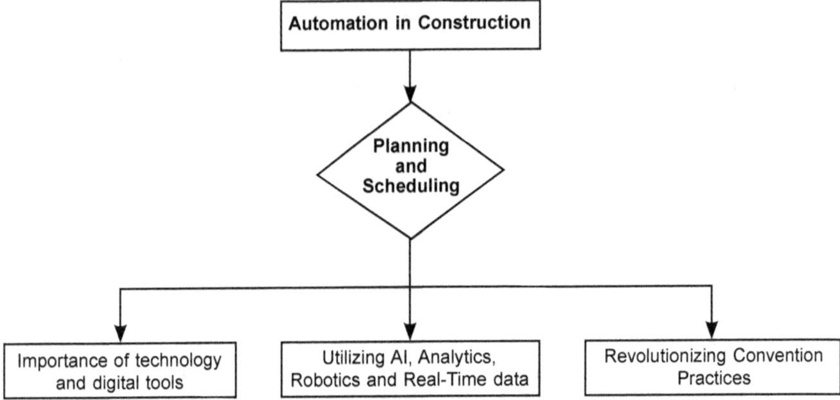

Figure 5.1: Automation in construction planning and scheduling.

Significance of Automation in Construction Planning and Scheduling

In the modern construction scene, automation in planning and scheduling plays a crucial role in changing traditional methods and advancing the sector's efficiency and competitiveness. The following essential points will help you understand the importance of automation:

- *Complexity Management:* Due to elaborate designs, changing rules, and involvement of several stakeholders, construction projects have grown more complex. Automation made possible by cutting-edge technology like BIM allows for improved collision detection, coordination, and visualization. Due to the ability to resolve possible conflicts early, design mistakes and rework during construction are decreased.
- *Time Management:* In building projects, efficiency is key. Cost overruns and poor client relationships can result from delays. Automation aids in developing reasonable and realistic project timeframes by generating optimized schedules based on historical data and real-time analytics. Automation reduces downtime, keeps the project on track, and assures timely project completion by automating the scheduling process.
- *Utilization of Resources:* Any building project must be successful in its use of resources. Based on the needs of the project and the resources that are available, automation technologies may intelligently distribute resources like labor, machinery, and supplies. This increases productivity by ensuring that

resources are assigned to jobs efficiently, maximizing resource utilization, and reducing waste.

- *Data-Driven Decision-Making:* Automation provides access to detailed project data and analytics to construction managers and decision-makers. This data-driven strategy enables risk assessment, performance evaluation, and informed decision-making. With the use of predictive analytics, construction teams can foresee potential delays and take proactive steps to reduce risks before they have an influence on the project.
- *Risk Management:* There are a number of intrinsic hazards associated with construction projects, including problems with the weather, a lack of materials, and unanticipated difficulties. Through simulations and scenario modeling, automation helps detect possible dangers at an early stage of preparation. Planning for contingencies is possible thanks to this proactive risk management strategy, which also lessens the effect of probable interruptions.
- *Cost Control:* Automation is essential for cost management and budget adherence. Automation aids in preserving project budgets by maximizing resource allocation, minimizing waste, and preventing delays. This not only lowers costs but also improves the project's overall financial sustainability.
- *Quality Control:* By reducing human error and guaranteeing adherence to project standards, automation helps to enhance the quality of construction. Robotics and prefabrication techniques are two examples of automated equipment that can improve accuracy and consistency in building operations, producing deliverables of greater quality.
- *Improved Collaboration and Communication:* Automation encourages effective collaboration and communication amongst project stakeholders. Cloud-based collaboration tools provide real-time data sharing, instant messaging, and document management. With this level of collaboration, there are fewer misunderstandings and better coordination since everyone is using the most recent information.
- *Sustainability and Green Building:* By maximizing material consumption, lowering waste, and improving energy efficiency, automation may help sustainable construction techniques. Construction teams can assess the environmental effects of design decisions and spot chances for sustainable building solutions thanks to BIM and automation.
- *Edge over Competitors:* Automation provides a considerable edge in the fiercely competitive construction sector. Automation may help construction businesses deliver projects more quickly, more effectively, and more affordably, bringing in more customers and building a reputation for excellence.

Automation is essential for scheduling and planning construction projects. Automation is a revolutionary force that helps construction organizations to flourish in a quickly changing market, from managing complexity and optimizing resources to allowing data-driven choices and boosting project quality. For construction companies seeking sustainable growth and success in the construction industry of the twenty-first century, embracing automation is more than simply a technological development—it is a strategic requirement.

Advantages of Automation in Construction Planning and Scheduling

Numerous advantages of automation in construction planning and scheduling revolutionize how construction projects are handled and carried out. These advantages boost cooperation, cost management, and overall project success in addition to increasing project efficiency and productivity. Using automation in construction planning and scheduling has the following main advantages:

- *Enhanced Efficiency:* Automation simplifies a number of project planning and scheduling procedures, saving time and paper. Automated workflows shorten the time it takes to make decisions and respond to requests, which speeds up the start and finish of projects.
- *Cost Savings:* For building projects, effective scheduling and optimal resource allocation lead to cost savings. Automation lessens idle time, eliminates resource overuse, and lowers the likelihood of costly project delays.
- *Accurate Scheduling:* Scheduling that is accurate is produced via automation, which takes job dependencies, resource availability, and external limitations into account. This results in timetables that are realistic, reducing the possibility of project delays and guaranteeing that projects are finished on time.
- *Improved Collaboration and Communication:* Automation makes it easier for varied project teams and stakeholders to collaborate and communicate in a seamless manner. Real-time data sharing and quick communication are made possible by cloud-based collaboration technologies, which improve coordination and decision-making.
- *Increased Productivity:* Productivity is increased because repetitive and time-consuming manual chores are eliminated via automation, allowing project teams to concentrate on important work. As a result, production rises at every phase of the project.
- *Resource Allocation Optimization:* Automation makes certain that the appropriate resources are assigned to the appropriate tasks at the appropriate times. As a result, disputes and resource shortages are avoided, resource utilization is increased, and waste is decreased, all of which improve project efficiency.
- *Enhanced Safety:* Automation of construction sites, including the use of robotics and drones, can eliminate the need for humans to do risky activities, enhancing site safety overall and reducing the likelihood of accidents.
- *Real-time Monitoring and Reporting:* Automation enables real-time monitoring of a project's progress and effectiveness. As a consequence, project managers may act quickly to intervene and make changes since they have the most recent information regarding the project's status.

- *Making Decisions Based on Data:* Automation gives users access to extensive project data and analytics. Project managers can make wise decisions, optimize procedures, and reduce risks thanks to data-driven decision-making.
- *Productivity Risk Management:* Risk management that is proactive is made possible by automation and its predictive analytics capabilities. Construction teams may put plans in place to reduce risks and prevent future project interruptions via early risk identification.
- *Quality Control:* Automation minimizes human error and guarantees adherence to project parameters, improving the caliber of construction and improving project results.
- *Sustainability and Green Building:* By maximizing material consumption, reducing waste, and assessing environmental implications, automation may help sustainable construction practices. This aids in the development of green and cost-effective construction solutions.
- *Competitive Advantage:* Adopting automation gives construction businesses a leg-up in the market. Automation enables businesses to complete projects more quickly, more effectively, and more affordably, bringing in more customers and establishing a reputation for excellence.

The advantages of automation in scheduling and planning building projects are numerous and wide-ranging. Automation helps construction businesses to optimize project performance, deliver projects more effectively, and remain ahead in a competitive sector through cost savings and improved resource management to greater cooperation and safety. Construction companies may live and compete in the dynamic and changing construction world of the twenty-first century by adopting automation, which is more than just a technology-driven choicc.

Automation in Construction: Quality Control and Safety

At its worst, construction rework can cause project delays, like at Berlin Airport, structural stability problems, as witnessed in Eindhoven Airport, or tragic consequences like those observable with scaffolding issues at Neurath power station. Reliable collection and review of field data are needed to mitigate the negative effects of construction rework. 3D point clouds provide a solid foundation for dimensional verification and tolerance checks during fabrication. In order to reduce the danger of disasters like structural failure and late discovery of rework, it is crucial to regularly collect and analyze 3D point clouds of building parts.

Quality Control

Wang created an integrated system using BIM and light detection and ranging (LiDAR) for on-site data collection and construction quality monitoring. It may assist quality managers in precisely and rapidly identifying and managing faults, which is an improvement over the time-consuming inspections that needed to be carried out at specific locations. To identify the possible use of laser scanned as-built models in quality assurance procedures of Liquefied Natural Gas (LNG) plant building, many case studies were undertaken by offering a situation awareness environment in the construction site. Other monitoring tools, like RFID, GPS, and so on, can also be utilized to quickly identify construction-related uncertainty and assist the manager in hastily revising their plans. By using cloud services, it is possible to build online communications between such an environment. To improve construction quality management and supervision, a system called the Construction Quality Supervision Collaboration System (CQSCS) was established (Jeong Kim et al., 2015).

Prefabricated Element Quality Control

Both *in-situ* and prefabricated construction can be subject to quality control utilizing point clouds. big modules are often manufactured in controlled manufacturing facilities before being brought to the site in order to prefabricate big megaprojects in remote or frigid areas. In the case of oil and gas projects, these modules must be connected at flanges after being delivered to the project site since they consist of multiple pipes and flanges. To efficiently verify the quality of flanges on these modules, a well-designed approach is necessary.

The previously described procedure offers a fresh approach to cylinder detection from point clouds for fabrication assurance. Along with the oil and gas plant example for module manufacturing verification, the strategy was effective in ensuring the quality oversight of anchoring nuts and bolts that are frequently utilized in prefab concrete and steel projects for construction (Maalek et al., 2021).

Automated Construction Control AI-based System

AI-based systems examine photos, videos, and sensor data taken during construction to automate quality control procedures. AI algorithms may detect flaws, faults, or deviations by comparing the data to preset criteria, guaranteeing that the built pieces adhere to the necessary quality standards (Dockery, 2023).

3D AI Construction System for Quality Control

The 3D AI Development Inspection system will provide automated fault identification, deployment quality authentication, and progress tracking for infrastructure and building construction projects. Asset owners, general contractors, and investors may now complete projects on schedule, under budget, and with minimal risk thanks to the solution.

Surprisingly, 60% of all construction projects encounter major delays, and the necessity for repair and installation problem correction can consume up to 30% of construction expenses. Construction projects may be completed more quickly, sustainably, and at lower prices thanks to digital techniques like BIM and Digital Twins.

Employing drones and commercially available Lidar scanners, the AI aids in autonomously monitoring progress on construction and deployment quality for both infrastructure and building projects. The ability to quickly spot any variations from the drawings or installation mistakes is provided by this automatic comparison, which may also assist maintain digital twins current throughout the building process.

The solution leverages cutting-edge 3D AI technology and 150 years of international testing and inspection experience to automatically provide useful information from 3D production site information (Lidar/Laser scans) instantaneously. This is done by automatically contrasting the related to structure, design, and fit-out elements of the 3D site data with the BIM and timetable after all those features have been intelligently recognized (Brown, 2021).

Automation Construction in Safety

AI has advanced quickly in recent years and now offers itself as a tool to boost productivity and safety in the construction sector.

At a jobsite, two construction workers are having a conversation about using AI to improve communication. Many construction firms are recognizing the operational benefits of AI and exploring methods to apply it to their safety and other departments.

Various processes may be automated using robotic process automation (RPA) and automation. This involves management tasks including scheduling, assigning tasks, and communicating. RPA may automate the job it needs to do by using robotic and digital technology and equipment.

AI and sensors may be used to track temperatures, find abnormalities, and more.

The Internet of Things (IoT) and AI infrastructure can communicate with one another. IoT links a variety of devices together throughout your digital network to move and synchronize data and manage data and information throughout your business (Rasheed et al., 2022).

Additionally, predictive analytics analyzes and monitors data. It may develop practical tactics using data sets and helpful insights. AI analytics can help you make informed decisions for upcoming projects, safety regulations, and policies.

AI Applications that Can Improve Safety

There are various applications of AI for construction industry safety. There are also useful ways to implement and benefit from AI for the security measures in your business.

Increase Automation of Risky Tasks

Robotic automation can assist in removing certain hazardous manual activities from the purview of your construction staff. An AI-supported robotic device is being used by two construction workers to assist automate a risky duty at the workplace.

By concentrating on less dangerous tasks, employees are less likely to be in danger of harm or injury. Like the Material Unit Lift Enhancer, which can perform heavy lifting of up to 6000 pounds per day, they can easily manage greater weights.

Robots with AI capabilities can also work more effectively and remove any potential safety issues brought on by a human error (Syed et al., 2020). For automation and self-learning, they employ clever algorithms and ML. Because of the way they acquire and evaluate data, they may collect data over time, learn from it, and get better over time.

This has the amazing benefit that AI robots can operate in hazardous environments where people may otherwise be exposed to health risks. This covers situations where manufacturing leads to the release of hazardous substances that might endanger the health and safety of workers (Bock, 2007).

Construction Accident Prevention or Reduction

The advantages AI can have for accident management and prevention go hand-in-hand with the decrease of risk-heavy jobs. One of the main causes of accidents in construction industries is human mistake that can result in catastrophic injuries and fatalities.

The U.S. Department of Labour reports that across all industries, accident rates rise by 18% and injury rates by 30% during evening and night shifts. This is relevant to the mental capacity of human workers and how it may affect their clarity, judgment, and productivity. Where robotics cannot be used by your company to handle dangerous labor, AI may be used to assess staff productivity.

Take into account AI-driven vision technologies, which have face recognition capabilities and can track facial expressions and detect signs of exhaustion like yawning, drooping eyelids, and more. On the workplace, a visibly worn-out construction worker is taking a break from his duties.

This can be helpful for a variety of employees at your organization, such as long-distance truckers transporting heavy equipment and supplies and on-site staff performing repetitive duties.

Determine Safety Risks and Anomalies

One use of AI as a safety detector is facial recognition technology. Another great technique to monitor employees' well-being and operational competence is through wearable sensors. When working with remote or lone employees, especially, pair gadgets like smartwatches with AI so they can warn you when an employee's vitals change.

AI anomaly detection can also spot health anomalies and help people take care of health problems early on. AI-powered sensors can also find anomalies in workplace safety. They can detect unfavorable temperature changes, behavioral variations, equipment malfunctions, possible risks, etc., and issue an early warning.

Protect Your Workplaces

It is challenging to ensure construction site safety since construction work is ad hoc. Contrary to office settings, where security elements may be permanently installed and monitored, construction sites cannot. You may build an infrastructure that improves construction site safety by combining several AI media. A security camera with AI assistance is scanning a building site for any unusual activity.

To build a network of security devices that watch over your site, combine the strength of AI vision, sensors, and the IoT. AI analytics, 360-degree cameras, and video can also use face recognition to detect people it doesn't recognize and warn you of possible invaders.

A 24/7 remote video surveillance system with AI integration can keep an eye on your property and deter theft, damage, and loitering. As you won't need to figure out how to safeguard your site and won't face project delays because of missing tools and materials or other problems, it also boosts your efficiency and capacity. Construction safety through AI and robotics is shown in Figure 5.2.

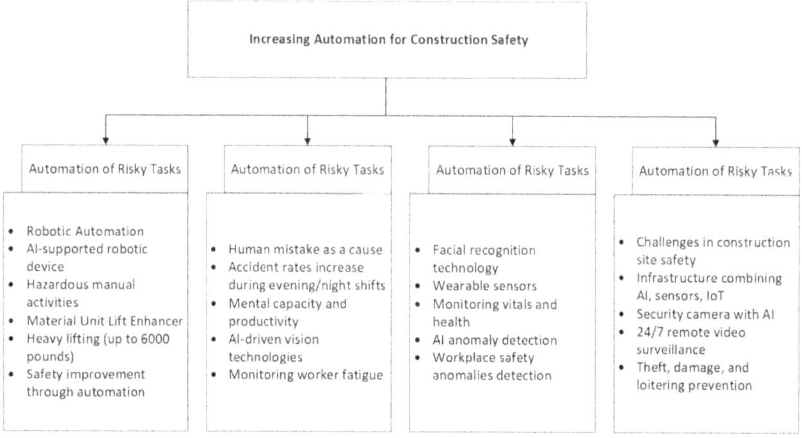

Figure 5.2: Enhancing construction safety through AI and robotics.

Optimizing Construction Safety: AI-Driven Efficiency

One of the key responsibilities of every construction firm is creating effective and thorough safety programs. It guarantees the health and safety of your staff and enables them to continue working effectively without worrying about these things. Putting safety first may help you avoid losses and delays brought on by accidents, deaths, theft, and other events. Adopt AI right now to improve the efficiency of your construction business (Choudhry et al., 2007).

References

Altaf, M., Alaloul, W.S., Musarat, M.A., Bukhari, H., Saad, S. and Syed Ammad. (2020). BIM Implication of Life Cycle Cost Analysis in Construction Project: A Systematic Review. *2020 Second International Sustainability and Resilience Conference: Technology and Innovation in Building Designs*. https://doi.org/10.1109/IEEECONF51154.2020.9319970.

Ammad, S., Alaloul, W.S., Saad, S., Qureshi, A.H., Sheikh, N., Ali, M. and Muhammad Altaf. (2020). Personal protective equipment in construction, accidents involved in construction infrastructure projects. *Solid State Technology*, 63(6). https://solidstatetechnology.us/index.php/JSST/article/view/3766.

Ammad, S., Alaloul, W.S., Saad, S., Altaf, M., Alawag, A.M. and Ali, M. (2021). Building Information Modelling (BIM) and Occupational Safety in Infrastructure Projects. *2021 International Conference on Data Analytics for Business and Industry (ICDABI)*, pp. 240–244. https://doi.org/10.1109/ICDABI53623.2021.9655832.

Ammad, S., Saad, S., Bashir, M.T., Qureshi, A.H., Altaf, M. and Rasheed, K. (2021). Construction Accidents via Integrating Building Information Modelling (BIM) with Emerging Digital Technologies: A Review. *2021 3rd International Sustainability and Resilience Conference: Climate Change*. https://doi.org/10.1109/IEEECONF53624.2021.9668066.

Bilal, M., Rasheed, K., Abbasi, M.F., Yamin, I., Khan, B.S.A. and Chowdhury, T.Z. (2022). Impact of Business Strategy on Project Management Elements Focus Moderating Role of Competition Attributes in Textile Industry. *2022 International Conference on Decision Aid Sciences and Applications, DASA 2022*. https://doi.org/10.1109/DASA54658.2022.9765222.

Bock, T. (2007). Construction robotics. *Autonomous Robots*, 22(3), 201–209. https://doi.org/10.1007/S10514-006-9008-5/METRICS.

Brown, A. (2021). *New AI Quality Control System for Construction.* https://constructiontechnology.media>news>

Burry, J., Sabin, J.E., Sheil, B. and Skavara, M. (2020). *Fabricate 2020*. UCL Press.

Chen, L., Li, S. and Tan, Y. (2023). Automated detection for the reserved rebars of bridge pile caps based on point cloud data and BIM. *In:* J. Li, W. Lu, Y. Peng, H. Yuan and D. Wang (Eds.), *Proceedings of the 27th International Symposium on Advancement of Construction Management and Real Estate*, pp. 1147–1162. Springer Nature Singapore.

Choudhry, R.M., Fang, D. and Mohamed, S. (2007). Developing a model of construction safety culture. *Journal of Management in Engineering*, 23(4), 207–212.

Davis, M. (2022). *Learn How Construction Automation Plays a Key Role in the Industry's Digital Transformation and in Solving the Global Housing Crisis*. https://www.autodesk.com>design>make>articles.

Debs, L., Zhang, J. and Chen, Y. (2021). *Introduction to Prefabrication and Automation in Construction*. https://docs.lib.purdue.edu>bcmoer.

Dockery, D. (2023). *AI in Construction has Landed*. https://www.constructconnect.com.

Jeong Kim, M., Chi, H.-L., Wang, X., Jeong, M., Chi, K.H.-L. and Ding, Lieyun. (2015). Automation and robotics in construction and civil engineering. Special Issue for Sensors "Intelligence Sensors and Sensing Spaces for Smart Home", View project Supporting Exports of International Creative Team's Services: Australia-Korea remote teamwork Vie. Article in *Journal of Intelligent & Robotic Systems*, 79, 347–350. https://doi.org/10.1007/s10846-015-0252-9.

Jiang, Y., Shu, J., Ye, J. and Zhao, W. (2023). Virtual trail assembly of prefabricated structures based on point cloud and BIM. *Automation in Construction*, 155, 105049. https://doi.org/https://doi.org/10.1016/j.autcon.2023.105049.

Jones, K. (2019). *Semi-automated Robot: Sam100 | PlanSwift.com*.

Khedmatgozar Dolati, S.S. and Mehrabi, A. (2021). Review of available systems and materials for splicing prestressed-precast concrete piles. *Structures*, 30, 850–865. https://doi.org/https://doi.org/10.1016/j.istruc.2021.01.029.

Lines, B.C., Sullivan, K.T., Hurtado, K.C. and Savicky, J. (2015). Planning in construction: Longitudinal study of pre-contract planning model demonstrates reduction in project cost and schedule growth. *International Journal of Construction Education and Research*, 11(1), 21–39. https://doi.org/10.1080/15578771.2013.872733.

Maalek, R., Lichti, D.D., Walker, R., Bhavnani, A. and Ruwanpura, J.Y. (2021). KIT - Our Departments - Digital Engineering and Construction - Research - Automated Construction Quality Control and Health Monitoring. *Automation in Construction*, 103, 150–167. https://doi.org/10.1016/J.AUTCON.2019.03.013.

Mohsen Alawag, A., Salah Alaloul, W., Liew, M.S., Ali Musarat, M., Baarimah, A.O., Saad, S. and Ammad, S. (2023). Critical success factors influencing total quality management in industrialised building system: A case of Malaysian construction industry. *Ain Shams Engineering Journal*, 14(2), 101877. https://doi.org/https://doi.org/10.1016/j.asej.2022.101877.

Muck, S. (2018). Concrete Focus: Meet the Rebar Robot. *Construction Management Magazine: Advanced Construction Robotics*. https://www.constructionrobots.com.

Naeem, U.H., Bilal, M., Syed, F., Rasheed, K. and Saad, S. (2022). A Multilayer Encryption Model to Protect Healthcare Data in Cloud Environment. *2022 International Conference on Data Analytics for Business and Industry, ICDABI 2022*. https://doi.org/10.1109/ICDABI56818.2022.10041708.

Pan, Y. and Zhang, L. (2021). Roles of artificial intelligence in construction engineering and management: A critical review and future trends. *Automation in Construction*, 122, 103517. https://doi.org/https://doi.org/10.1016/j.autcon.2020.103517.

Rasheed, K., Ammad, S., Said, A.Y., Balbehaith, M., Oad, V.K. and Khan, A. (2022). Application of Smart IoT Technology in Project Management Scenarios. *2022 International Conference on Data Analytics for Business and Industry, ICDABI 2022*. https://doi.org/10.1109/ICDABI56818.2022.10041674.

Saad, S., Alaloul, W.S., Ammad, S., Qureshi, A.H., Altaf, M. and Rasheed, K. (2020). Design Phase Carbon Emission Prediction Using a Visual Programming Technique. *2020 2nd International Sustainability and Resilience Conference: Technology and Innovation in Building Designs*. https://doi.org/10.1109/IEEECONF51154.2020.9319978.

Saad, S., Alaloul, W.S., Ammad, S. and Qureshi, A.H. (2022). A qualitative conceptual framework to tackle skill shortages in offsite construction industry: A scientometric approach. *Engineering, Construction and Architectural Management*, 29(10), 3917–3947. https://doi.org/10.1108/ECAM-04-2021-0287.

Sánchez, F. and Hartlieb, P. (2020). Innovation in the mining industry: Technological trends and a case study of the challenges of disruptive innovation. *Mining, Metallurgy & Exploration*, 37(5), 1385–1399. https://doi.org/10.1007/s42461-020-00262-1.

Solomom, V. (2023). *8 Benefits of Automation in Construction – Rossum*. https://rossum.ai/blog/automation-in-construction/

Sweet, R. (2018). The contractor who invented a construction robot. *Construction Research and Innovation*, 9(1), 9–12. https://doi.org/10.1080/20450249.2018.1442702.

Syed, R., Suriadi, S., Adams, M., Bandara, W., Leemans, S.J.J., Ouyang, C., ter Hofstede, A.H.M., van de Weerd, I., Wynn, M.T. and Reijers, H.A. (2020). Robotic process automation: Contemporary themes and challenges. *Computers in Industry*, 115, 103162. https://doi.org/https://doi.org/10.1016/j.compind.2019.103162.

Tajenkar, Akshay S. (2017). *The Use of Automation in Construction Industry – Civil Engineering Portal – Biggest Civil Engineering Information Sharing Website*. https://www.engineeringcivil.com/the-use-of-automation-in-construction-industry.html

Wang, M., Wang, C.C., Sepasgozar, S. and Zlatanova, S. (2020). A systematic review of digital technology adoption in off-site construction: Current status and future direction towards Industry 4.0. *Buildings*, 10(11). https://doi.org/10.3390/buildings10110204.

Wang, W.-C., Weng, S.-W., Wang, S.-H. and Chen, C.-Y. (2014). Integrating building information models with construction process simulations for project scheduling support. *Automation in Construction*, 37, 68–80. https://doi.org/https://doi.org/10.1016/j.autcon.2013.10.009.

Wu, P., Wang, J. and Wang, X. (2016). A critical review of the use of 3-D printing in the construction industry. *Automation in Construction*, 68, 21–31. https://doi.org/https://doi.org/10.1016/j.autcon.2016.04.005.

CHAPTER

6

Robotics in Construction

Syed Saad, Ahmad Zaland, Syed Ammad and Kumeel Rasheed

Introduction

Since the start of the twentieth century, the usage of robots in construction and architecture has grown significantly, and they are increasingly pushing the boundaries in both industries. Numerous sectors are interested in the cutting-edge subject of robotics. In commerce, the military, home chores, architectural design, and construction, robots are now often used. The development of robotic systems has drastically changed the available design methodologies and has begun to push the boundaries of architecture, a subject of intense study by architects and designers. Industrial robots may be programmed by architects with little to no programming knowledge using specialized Grasshopper3D plug-ins like KUKA|prc or HAL. Digital parametric settings and robot codes may both be changed in a matter of seconds. Procedures are moving forward swiftly; this development in digital fabrication fundamentally changes how architects interact with technological devices and offers a new perspective on the interaction between the digital and physical worlds, enabling the modification and individualization of architectural design and its materialization with minimal time and effort.

The International Organization for Standardization (ISO 8373) defines a robot as a device that is automatically operated, can be reprogrammed, and can manipulate programs in an industrial context (Stone, 2018). A number of operational features describe the construction sector. It involves heavy lifting, needs great precision, is a dangerous work, and time management is critical. Most construction-related tasks were either carried out manually or relied on machinery that needed human control up until relatively recently. However, because manual labor was insufficient, automated and semi-automated technologies were created. Robotics are now used in almost every area of the construction business, according to Balaguer et al. (2002): demolition, surveying, excavation and earth movement, tunneling, paving, home building, and inspection tasks. Each of these activities has certain tasks that call for using particular robots or automated equipment.

Brief History of Robots in Construction

The first construction robots were developed in the early 1970s to enhance the standard of prefabricated modular homes in Japan, and in the late 1970s, preparations for their use on construction sites started. In the 1980s, the first construction robots appeared on job sites. By the 1990s, integrated automated building construction sites had been built and employed around 20 times more robots. Additionally, building security robots as well as maintenance robots for cleaning and inspecting infrastructure, buildings, and real estate have been developed. Humanoid construction robots had been tried throughout the first ten years of the twenty-first century. Service robots will have a significant market in the built environment in the future (Bock, 2007).

Types of Robots Used in Construction

There are many different types and sizes of robots. Robots cannot be used in construction. It necessitates the use of heavy-duty robots that can operate independently in severe conditions without the need for a significant amount of human labor. This article will discuss some of the many robot kinds that are employed to carry out particular building jobs.

Demolition Robot

In particular for any remodeling site, demolition is a necessary step in the building process. Large-scale building and structural destruction may be speeded up, made more efficient, and cost- and time-effective by using robots (Babbar et al., 2021). It makes it safer for human employees since it enables the operator to work while keeping a safe distance from toxins, debris, and crumbling concrete. In constrained spaces, robots make it simple to knock down walls, gather trash, and smash concrete.

Robotic Bricklayers

The most repetitive and laborious operation in building is placing bricks. The task's repetitiveness raises the chance of mistake. However, the bricklaying robots allow us to swiftly do such boring chores.

Notably, the "Motor Mason", the first robotic bricklayer, received a patent in 1904. It resembled the bricklaying robot SAM (Semi-Automated Mason), which was created in the United States by Construction Robotics in collaboration with the National Science Foundation. Three workers were needed to operate the "Motor Mason". It was thought to be five to ten times quicker than a human.

While other human-operated robots mix cement or bonding agent, most bricklaying robots use the arms of industrial robots to assemble the masonry framework of the building. Efficiency and quicker construction are the main gains. The installation costs are the primary drawback to this technology, and the robots are only useful for massive building projects with short deadlines. It is noteworthy that SAM can lay 3000 bricks in an eight-hour shift.

Robotic Welders

Due to the lengthy setup process, pain experienced by the operator, safety concerns, and expense, manual welding is only effective for brief periods of time (Hägele et al., 2016). Therefore, robotic welding is essential to construction in order to produce high-quality welds in less time. The majority of human element flaws are eliminated by robots, improving productivity, quality, and cost-effectiveness. They can manoeuver with ease in tight spaces and weld precisely.

Robots that weld pose certain risks despite the benefits. A typical firm cannot afford one since they are expensive to obtain. They frequently need maintenance or repair and need to be operated and programmed by qualified persons. They are also restricted to a small number of welding techniques, many of which need more time to cool or, when applied improperly, may even weaken metal.

Exoskeletons

Another cutting-edge robotics innovation utilized in building is the exoskeleton. A typical worker's strength, speed, and agility may be greatly increased by this mechanical outfit. Any worker can lift and move heavier things than the usual person thanks to it. Here are some instances of exosuits being utilized in building:

- *Mounted Arm Exosuit:* This is a tool-holding suit with a spring-loaded arm that can support a lot of weight. They facilitate the worker's usage of heavy equipment and contribute to a quicker, less-fatiguing job completion (K. et al., 2018). The back of the worker is supported by this exosuit as they repeatedly bend down and carry large weights. It makes sure the weights are handled properly to prevent back injury. When doing such repetitive lifting chores, they lessen the tension placed on the back muscles.
- *Arm Support Limb:* By supporting the shoulder and arm, this outfit aids workers in lifting heavy tools. When using heavy tools for an extended period of time, it reduces the strain and stress above the waist.
- *Support for Crouching and Standing:* These portable chairs help the user maintain a crouching or standing position for an extended period of time. They provide the same kind of support as a chair for the user.
- *Whole-Body Suits:* This style of suit maintains the posture of the full body and offers additional strength. The bulkiness and general lack of speed and

agility of this technology are drawbacks. Exoskeletons are typically pricy, heavy, unwieldy, and difficult to wear for extended periods of time due to the utilization of electric motors and bulky materials.

Contour Crafting and 3D Printing

The broad use of 3D technology over the past three decades has completely changed the manufacturing industry, enabling everything from rapid prototypes to fully working cars (Qureshi, Alaloul, Murtiyoso, et al., 2022). It has just entered the construction industry and may be used for a variety of tasks, like building homes in a single day or an apartment complex in a week.

This technology has the potential to change the fundamental process of building because of its capacity to boost production and reduce waste. With the use of 3D technology, a work that would need months to accomplish using the conventional building technique may now be finished in a day or two (Ammad et al., 2021). If the printer is given the proper specs, it can create a broad range of items. Additionally, it lowers expenses and boosts productivity.

A layered manufacturing technique with a lot of promise for automated construction is contour crafting. It is possible to create a single house or even a whole estate of homes in a single run, with each one having a unique design. Contour crafting offers potential in two areas: (1) low-income housing or emergency sheltered housing; and (2) architectural projects incorporating complicated shapes that would be expensive to construct using conventional techniques. This is because of its quickness and capacity to employ *in-situ* materials.

Drones

Another recent development in building technology is this one. Drones are essentially autonomous robots that operators remotely command to complete a variety of difficult jobs, such as creating contour maps in three dimensions, scanning objects in three dimensions, moving objects, conducting surveys, keeping an eye on the sky, measuring volumes (Hussein et al., 2021).

Drones enhance site safety, enhance participant communication, and save project time and expense. They may use the construction elements to provide real-time aerial views that show the terrain's advantages and disadvantages. Architectural engineering, construction management, and monitoring operations are now more realistic and cost-effective because of recent technology advancements in low-weight, autonomous drone design, and navigation (Qureshi, Alaloul, Wing, et al., 2022). The labor-intensive nature of conventional construction monitoring and reporting methods may be greatly reduced by drones. They can offer practical and clever methods for managing and supervising construction sites, which can lead to improved operations, better planning, and more efficient on-site modifications as shown in Figure 6.1.

Figure 6.1: AI generated image of drone collecting real-time data on construction site.

Human-like Robots

The traditional picture that comes to mind while thinking of robots is of humanoid workers. They are now more real than imagined, though. Despite of what was previously said, there is a labor shortage in the construction industry since few individuals are interested in pursuing careers in the industry Now, the robots.

The closest thing we have to androids is the HRP-5P created by the National Institute of Advanced Industrial Science and Technology in Japan. By fusing environment sensing with object identification technology, it is capable of a wide range of activities. It already has the ability to do many construction tasks including erecting plasterboard sheets and utilizing power tools—while still at the prototype stage.

Advantages of Robotics in Construction

It is apparent that there are several robotics efforts, each of which is progressing independently. We shall analyze the following categories of automation and robotic technology in an effort to focus a broad field:

- Prefabrication off-site
- Localized robotic and automated systems
- Drones and self-driving cars
- Exoskeletons.

It's also important to keep in mind that the "sense technology" in the majority of these situations is provided by LiDAR and point cloud technology.

Assisting in Addressing the Construction Industry's Skills Gap

The construction industry requires a lot of workers. Automating tasks with robots has demonstrated a high degree of effectiveness in lowering labor costs while also increasing quality and productivity in other industries.

There have long been worries about a worker shortage in the building industry. Productivity can be increased, and this shortfall may be reduced by using robots. Additionally, it will enable workers with superior talents to earn higher compensation. Those with advanced talents will undoubtedly be in higher demand, at least throughout the transition period, which will last for at least a decade.

This essentially means that there will be up to 200 million more construction jobs than there are now by the year 2030. Increasing automation in the construction process will hasten the delivery of structures and infrastructure without needing to reduce manpower. Robots will reduce skill shortage issues, but they won't replace people and they will spur the need for new kinds of skilled jobs (Saad et al., 2022).

Utilizing Off-site Production to Accelerate the Process

Compared to what can be done on-site, manufacturing individual components or modules in factories lends itself to increased automation. Building construction might be significantly impacted by a large move to off-site modular construction, but the change will take time.

Such modules are already being constructed by some businesses, including Katerra. The majority of the building work in these factories is done manually, but as the scale grows, the process will become increasingly automated.

High-level construction components are being combined into full building modules such as bathroom or kitchen modules, using large-scale prefabrication (LSP) techniques (Davila Delgado et al., 2019). The use of additive manufacturing methods like 3D printing, which are already having an influence and producing examples like 2-storey municipal buildings and 3D printed bridges, will also be advantageous in this.

Assessments of a site and its surroundings will be required before to, during, and following a construction project to guarantee accuracy (Saad et al., 2020). The cost of doing these thorough point cloud scans has decreased because of developments in point cloud registration (related to algorithm upgrades and cloud-based registration).

Construction sites or prefab supplies may be scanned and cross-referenced with plans thanks to scan-to-BIM and the related point cloud registration. This will enable the construction operations to be checked for quality.

According to McKinsey, 15 to 20% of new building construction will be modular by 2030. It's a gradual uptake, and many activities will continue there for some time. However, it is a major change, and because of the effectiveness and cost savings it provides to building, its use will only increase.

Improving the Effectiveness of Sites

On construction sites all throughout the world, automated and robotic technologies are being tested and piloted. With different degrees of accomplishment, tasks including laying bricks, assembling steel trusses, welding, installing, painting, and pouring concrete are all being mechanized.

These devices are referred to as single-task construction robots (STCRs), which carry out a single task repeatedly (Melenbrink et al., 2020). Hadrian X, a robotic arm used for bricklaying, is a classic illustration.

Until now, the challenge has been getting these specialized robots to cooperate in the hectic atmosphere of a building site. Robotic on-site factories have been proposed as one way to standardize the environment somewhat, although they are still in their infancy and may cause more issues than they resolve.

It's doubtful that you'll dismiss a bricklayer and hire a robot to perform what the bricklayer did. It's more likely that machines will replace certain tasks inside a position. Workers will thus need to have the skills necessary to coexist together with robots or to do hybrid tasks. Even the typical construction worker will eventually use a tablet to examine building designs or fly a drone instead of physically touring the job site.

Regular inspections will also need to be carried out by project managers and supervisors. Construction robotics and drone-produced 3D point clouds will be useful in these new procedures, especially when coordinated with BIM-enabled planning.

Creating Safer Working Conditions

One of the most important duties of a construction manager is to provide a safe working environment. A dangerous site has a lot of expenses in addition to the worries about preventing worker injuries. Robots are being considered as a way to increase safety in some of the riskier jobs, such as demolition.

Exoskeleton utilization, which was initially pioneered by the military, has expanded from the healthcare sector to include manufacturing and construction (Qiu et al., 2023). By allowing older people to continue working on-site and doing physically demanding activities, they can also be a solution to the problems caused by an ageing construction workforce.

Automating Drilling, Digging, and Moving Dirt

The employment of autonomous trucks and excavators for mining has so far

Table 6.1: Impact of automation and robotic technology on the construction industry

Categories of automation and robotic technology	Key points
Prefabrication off-site	Automation of building materials production in controlled settings. Enhanced quality and reduced on-site assembly time.
Localized robotic and automated systems	Robots used for various construction tasks, such as bricklaying, welding, and concrete pouring. Specialized robots for single tasks (STCRs).
Drones and self-driving cars	Drones used for site surveys, monitoring, and coordination. Potential for self-driving cars in construction logistics.
Exoskeletons	Exoskeletons aid workers in physically demanding tasks. Can address challenges of an aging construction workforce.
Impact on addressing the construction industry's skills gap	Automation can increase productivity and reduce labor costs. Anticipated worker shortage in the industry. Robots may create new skilled job opportunities.
Utilizing off-site production to accelerate the process	Modular construction in controlled factory settings. Additive manufacturing (e.g. 3D printing) for components. Use of point cloud scans and Scan-to-BIM for quality checks.
Improving the effectiveness of sites	Mechanization of tasks like bricklaying and welding. Challenges in getting specialized robots to cooperate. Workers may need to coexist with robots and perform hybrid tasks.
Creating safer working conditions	Robots considered for risky tasks like demolition. Exoskeletons to aid in worker safety and physical demands.
Automating drilling, digging, and moving dirt	Use of autonomous trucks and excavators, especially in remote construction sites. Automation in earthmoving equipment.
Entering perilous and harsh places	Drones for safety applications, inspection, and coordination. Challenges in ensuring the safety of robotics. Importance of LiDAR and point clouds for navigation.

been the most prominent way that robots have been used in large-scale building projects as illustrated in Table 6.1. The automation of earthmoving equipment still faces several obstacles on conventional building sites (Ammad et al., 2020).

For instance, Built Robotics has concentrated on projects at remote construction sites far from human laborers, such as roadway paving. The ground is prepared for subsequent stages of building when human personnel arrive by autonomous heavy machinery.

Entering Perilous and Harsh Places

Drones are playing more and more important roles in a variety of safety applications, from inspection to fire and security. Drones can monitor the status of a project and may ultimately help to coordinate the use of heavy machinery, equipment, and laborers on 'hybrid' construction sites (Li & Liu, 2019).

Robotics' safety features have potential, but there are still a lot of issues that need to be solved. It is getting easier to navigate. LiDAR, point clouds, and other reality capture technologies are becoming increasingly important in building far more accurate 3D models that the robots may use for navigation. But the system is still far from ideal.

Point clouds can give useful and dynamic site and location information (Naeem et al., 2022), especially when used with BIM. These 3D representations are essential for connecting robotic technology to the real world.

Case Studies of Robotics in Construction

Case Studies on Glazing Robot Technology on Construction Sites

A hybrid motion typed curtain-wall glazing robot (HCGR) has been designed and implemented to actual building sites to specifically meet the demands for handling curtain-walls in a precise and safe manner (Thompson, 2014). A macro-micro motion manipulator is the designed robotic system overview. The macro motion manipulator for lifting and moving curtain walls is referred to as a mini-excavator. The newly created three-degree-of-freedom robotic arm is regarded as a micro-motion manipulator for careful manipulation of curtain walls. Additionally, a unique control approach is proposed for a human-robot cooperative construction method in which an operator may manage construction materials intuitively. The HCGR prototype undergoes field tests on actual building sites. A comparison of the prototype robot's and the human-based building method's respective productivity is carried out.

Case: Hybrid Motioned Curtain-wall Glazing Robot (HCGR)

The research and creation of novel construction materials have advanced along with the trend towards larger and higher structures and buildings (Saad et al., 2021). Building construction professionals are particularly interested in curtain-walls since they are crucial to a building's adiabatic performance, water tightness, and aesthetics. But because there isn't enough construction equipment for handling curtain walls, the procedure is difficult and dangerous, requiring a lot of labor and the utilization of already-existing machinery. A mini-excavator-based automated construction is needed to address these issues. The tiny excavator method makes it simple to transfer a curtain wall to the assembly location. Additionally, this approach cuts down on labor costs and building time. However, a construction person must still operate the curtain-wall assembly. Therefore, as an example of an appropriate robotic system for carrying out a straightforward and secure curtain-wall handling approach, we suggested the use of "a Hybrid motioned Curtain-wall Glazing Robot (HCGR)". A 3 DOF robotic manipulator is part of the proposed system. Robotic trajectories are produced by the manipulator under human control.

Motion Manipulator

A manipulator for macro-micro motion is the HCGR overview. The macro-motion manipulator is referred to as a small excavator. A micro-motion manipulator is thought to be the 3 DOF robotic manipulator. A manipulator for macro-micro motion is the HCGR overview (Lee & Moon, 2015). The macro-motion manipulator is referred to as a small excavator. A micro-motion manipulator is thought to be the 3 DOF robotic manipulator. The micro-motion manipulator is useful for different types of construction work in addition to installing curtain walls. Therefore, this manipulator is modularized to add or subtract DOF as target building projects progress. The 3-DOF modularized manipulator and the finished HCGR's form. Each module is joined in a sequential sequence to represent the chosen DOF, and the order is then strongly tied to the system's operational effectiveness and safety. It is therefore illogical for an actuator to provide more force and torque than is required. After a worker used more power than was necessary to operate the manipulator, this would cause the entire work capability to decrease.

Control Strategy

Due to the continuously shifting surroundings associated with construction, a completely automated system is not appropriate for the task. Therefore, a cooperative human-robot system is appropriate for building tasks. In order to work with people, it is an interactive system. In general, a robot can move faster and more forcefully than a person who is sluggish, uses little energy, and frequently makes mistakes. On the other hand, the human mind, body, and conduct are considerably more adaptable and pliable. We state that this is because integrating human and robot advantages into cooperative human-robot manipulation would increase the system's effectiveness or performance from the perspectives of human error and labor productivity. The design of the robot controller takes into account interactions between the operator, robot, and environment. By applying operational force with a certain power assist ratio, the system, to which the proposed control technique is applied, enables an operator to handle building materials as if they were handled by himself. Additionally, by allowing an operator to sense environmental reaction pressures while doing an operation, this technology helps them do it more naturally. To implement the human-robot cooperative manipulation, an intelligent HRI (Human-Robot Interface) device is introduced. An operator can control the robot with sensor A (6 DOF force/torque sensor) by applying an external force that contains an operation command to the handler of the robot controller (Mathiassen et al., 2016). Here, sensor B (6 DOF force/torque sensor) transmits data on the contact force to the robot controller in the event that the robot makes contact with an outside object. The external force that is sent to sensor B and the external force that is transmitted to sensor A should function independently of one another method of HCGR regulation. A 3-DOF robotic manipulator is operated by the worker using an HRI device that is placed between the manipulator and a curtain wall (Manuel Davila Delgado & Oyedele, 2022). A control algorithm creates a command signal based on the data from this controller and sends it to the motion controller board through a digital analogue (D/A) converter. The necessary control is then accomplished after the encoder attached to the AC servo motor provides real-time input on how things are operating at the moment.

Field Test

The current handling approach (which relies on people) used on a real building site is examined and contrasted with the robotic handling method in order to gauge the proposed robotic system's productivity. By using the suggested robot on a building site, the curtain-wall handling method is shown. When evaluating the productivity of various handling techniques, the operators' opinions might be quite helpful. A survey that was distributed to each curtain-wall installer produced results that we acquired. They were questioned on their existing procedure for managing curtain walls, their expectations for the future, and the elements they felt were crucial to the robot's success. This helped us to pinpoint areas for robot development enhancement.

Case: Glass Ceiling Glazing Robot (GCGR)

The current glass ceiling glazing procedure, which depends on a scaffold (or aerial lift) and human labor, is risky and complex. Operators are at risk of falls, vehicle rollovers, and other incidents due to this process. Additionally, poor working postures, which can lead to a variety of musculoskeletal illnesses and reduce focus while working, have a significant role in increasing the incidence of accidents. That is to say, it contributes to a decline in construction productivity and safety. Poor working postures have been regarded by many ergonomists and health and safety professionals as one of the primary causes of musculoskeletal problems on industrial sites ever since (Valero et al., 2016). It is noted the detrimental effects for forceful and irregular motions and unnatural postures of the body, the building's soffit and the location of the glass ceiling glazing. Building dimensions are 32 m × 22 m, and glass ceiling glazing heights are 7.9 m and 15 m. Glass ceilings may be divided into three kinds for handling. The 750 mm × 1500 mm and 40 kg glass ceilings fall under the first group. The glass ceiling in the second category is 1500 mm × 1500 mm and weighs 80 kg. The glass ceiling in the final category measures 3000 mm × 1500 mm and weighs 150 kg. We described "Module T&H-bar" as an installation technique that is depicted as 'Lay-in' to install glass ceilings on ceiling fames.

Robot Hardware

The primary purpose of the glass ceiling glazing robot is determined by spot analysis and currently used techniques. To install large-sized ceiling glass securely, special equipment that can work with many workers is first needed. To ensure consistent construction quality, the methodology required for the worker's technology might also be reflected as is throughout installation. It is necessary to create the best installation plan design for higher productivity and an antidote to prevent mishaps using simulations and spot testing. The idea design of a GCGR is as follows, in accordance with the primary roles identified by work analysis.

1. It is necessary to have an aerial lift with enough operating range to support an operator and the installation equipment at a height of around 15 meters above the ground.
2. To install the bulky glass ceiling, a multi-DOF manipulator with several operators is required. The operating area and payload must be taken into consideration while selecting the robot.
3. To adapt to a changing work environment, this system is semi-automated. To operate a multi-DOF manipulator, one operator climbs onto the aerial lift's platform. The higher robot controller is now the worker's discretion.
4. External force information is employed as the input signal for the robot exercise in order to reflect the worker's capability. According to the intended robot exercise and the operator's age, the force information that utilizes the input signal must be allowed to vary freely.
5. To manage the glass ceiling, a hoover suction device is employed as an end-effect device. A modulation design process is used to create an end-effector in anticipation of variations in glass ceiling shape.
6. The worker and robot are both supported by the aerial lift's deck. The worker's safety and productivity must be taken into account in the design of the deck and the work procedure.

Hardware and software categories are used to group the core tasks identified through work analysis. The basic system is recommended to employ an industrial multi-DOF manipulator and an aerial lift. By matching the operator with a manipulator, the HRI gadget is involved in installing glass ceilings (Hersh, 2015). The operator's intentions are transmitted to the robot controller via this gadget. It is made up of two 6-axis F/T sensors and is situated between the vacuum suction device and the multi-DOF manipulator's flange.

Robot Control

While motion under contact circumstances requires accurate motion with a very slow robotic action, free space motion—motion between glass ceilings and glazing frames—needs quick movement with relatively low accuracy. We created an impedance controller for human-robot collaboration based on simulations of interactions between the operator, robot, and environment. Operator force is provided to sensor A when the operator determines that the location (X) to which a robot is carrying glass ceilings does not coincide with the position (Xd) to which he or she wishes to move them. Operators of varied ages can employ the external human force (Fh) detected by sensor A through the force augmentation ratio (α). Equation (1) for impedance gives the goal dynamics required for operation in terms of an operator's inputted force and the contact force (Fe) with surroundings inputted from sensor B. As input is received through the encoder of a position/direction controller, the dynamics values' discrepancy between the goal position (X_d) and the current position (X) diminishes, resulting in a value of 0. In other words, a servo controller receives the current deviation and instructs a manipulator to seek the desired position value. By adjusting the impedance parameters (M_t, B_t) in equation (1), it is also possible to modify the motion characteristics of a robot. Controlling these factors enables the implementation of relatively quick and accurate movements.

$$X_d = (M_t)^{-1} \{(\alpha F_h - F_e) - B_t X_d''\} \quad (1)$$

X_d : Acceleration related target dynamics

X_d'': Velocity related target dynamics

M_t : Inertia related impedance parameter in the virtual system

B_t : Damping related impedance parameter in the virtual system

Field Test

The suggested robotic system is designed to build the outside (glass ceiling) soffit. The installation distance for glass ceilings is 15 meters, and their dimensions are 3000 mm × 1500 mm and 150 kg, respectively. The suggested handling method may be summarized as follows based on the job planning:

1. The installation of an aerial work platform with a mounting platform for the proposed robotic system.

2. Filling the deck's glass ceiling.
3. Raise the glass ceiling close to the frame where it is installed.
4. Put up the glass ceiling and complete the project.

One piece of glass from the ceiling is loaded onto the deck before it is raised. The robot will then move towards the installation place in accordance with the deployment strategy. The comparison and analysis that results can be modified to fit the working conditions of the actual building site. Smaller structures require more workers when managing construction materials (Miozzo & Dewick, 2002). But given the inclination of current building trends towards bigger and higher structures, the GCGR's provided projections here offer a positive prognosis.

We must carry out more work necessary for application before we can apply human-robot cooperative manipulation at actual construction sites. First, it is determined that the conceptual design of a construction robot for handling building materials is based on an examination of job definition and working conditions. After that, a description of the robot's detailed design based on its field robot design follows. After a real-world field test on a building site, the designed system's productivity and safety are then contrasted with those of the existing glazing machinery. A transportable platform from the HCGR and GCGR is paired with a robotic manipulator to accommodate different working environments and building materials. As a result, different construction sites may manage a range of building materials (Lee & Moon, 2015).

Case Study: Robotic Cell for Construction Scotland Innovation Centre

A robotic cell with a range of end effectors was needed by the Construction Scotland Innovation Centre (CSIC) to develop and test various production processes.

There is a strong drive within the industry to execute as much of the work as possible in the factory rather than on-site in order to eliminate the health and safety concerns associated with handling large windows on-site.

We offered a modular and configurable cell with the following integrated components' design, production, installation, and initial commissioning:

- Robot KUKAI KR210 R2700
- Quantec Safe Operation to optimize the cell's size while guaranteeing that the robot cannot operate outside the cell is used for guarding
- Changes tools automatically
- PLC control system
- Safety mechanisms and apparatus connected to safety
- Software for system control that makes it simple to create programs, choose programs, sequence cells, and monitor safety
- End-effector robots

- End effector for the vision system
- Gripper end effector for windows
- End effector for nails
- End effector for a slot cutter
- End effector for a machining spindle
- Web application and vision software toolkit for configuring the vision system.

End Effector for the Vision System

An end effector equipped with a black-and-white camera to capture images. It functions in conjunction with the vision software toolset from Loop Technology, which enables the reading of barcodes and the location of corners and rounds with respect to a specified point. This enables CSIC to program the robot to find an object or feature of interest's precise location on its own. They cannot be at their current places in the cell without it. For instance, to precisely position a window, the window and its placement target must be lined up in the exact same location each time. The vision system offers the crucial adaptability needed for a very flexible workflow.

Gripper End Effector for Windows

A pick and place end effector that is 2.6 meters long by 1.5 meters broad and has adequate contact and adhesion to handle up to 120 kg. It is made out of a manually adjustable set of suction cups that can be moved in all directions for a totally customizable setup, including Z axis modifications. This enables CSIC to manage a variety of windows, including those that protrude from or are recessed into the same frame.

Inserting a Nail

An end effector having consecutive triggers for controlling hand-held power tools.

To fulfill a tool safety feature that needs engagement before firing, it uses a solenoid to pull a finger trigger and a modular carriage mechanism to depress the nose. This enables CSIC to automate the usage of several hand-held power tools, including staplers and nailers, in a secure manner.

Slot Cutter

An angle grinder end effector. Since it produces the necessary kerf, CSIC may utilize it as a slot cutter to make slots in wooden panels.

Cutting-edge Spindle

A milling end effector equipped with a high-speed spindle. As a result, CSIC is able to carry out machining with a big working envelope.

Challenges

The rebound from the nailer sent forces into the robot that affected the location of the tool was one of the project's numerous difficulties. These forces, which were not only challenging to measure precisely but also extremely varied and reliant on the hardness of each material, were challenging to account for.

To solve this, a custom counter force sliding rail was created, which had no bearing on the robot. In order to lower the cost of developing and installing new tools, it was also built to support a variety of interchangeable hand tools.

The overall issue of making hand tools safe for automated usage presented another difficulty. This requires meticulous tool safety zone determination using KUKA Safe Operation, localized guarding when necessary, and a detailed risk assessment to establish appropriate risk management measures.

Robotics in Construction: Site Surveying and Mapping

The construction sector is increasingly using robotic devices, both on the ground and in the air. The developed Robotized Construction Site Surveyor (RCSS) is the result of the fusion and task-specific specialization of research conducted across domains and applications (Jo et al., 2019), including those of nuclear site characterization, underground exploration, industrial inspection, navigation in low-visibility environments, such as dust and smoke, agile control, and more. This application-specific robotic platform was created by modifying and integrating the various techniques and systems created for such jobs. The resultant technology has undergone testing and verification in a variety of field conditions, including intricate building environments and a public housing construction site.

Characterization of Construction Sites in Multi-modal Form

Multi-modal sensor fusion for localization, mapping, object recognition, and position reporting is a crucial part of the Robotized Construction Site Surveyor. Building modeling requires accurate mapping of construction sites, which is accomplished by RCSS through the integration of LiDAR point clouds, visible-light and thermal vision cameras, and inertial measurement cues (Madonna et al., 2018). The purpose of this multi-modal sensor fusion is to achieve two objectives: (1) enable accurate mapping even in the worst conditions (such as at night or when there is a lot of dust and other obscurants), and (2) deliver a thorough mapping result that includes structural 3D information along with vision-based texture data and thermal signature information. The latter is used to identify construction flaws or pre-existing issues with a building (for instance, in the event of a rehabilitation project). Additionally, in order to assist with inventory keeping and maintenance, these robot skills have been expanded to include the capacity to recognize objects of interest and provide their specific position on the map.

A loosely connected method that integrates LiDAR-based odometry and mapping with visible-light and longwave infrared camera-inertial updates facilitates the process of multi-modal localization and mapping. High update rates (such 20FPS or 30FPS) are offered for camera-inertial updates. In most locations with adequate lighting, visible-light cameras produce accurate findings, but they struggle in conditions of darkness, lack of texture, or the presence of obscurants (such as dust, which is frequently present in construction sites). For camea-IMU based estimate, the Robust Visual-Inertial Odometry (ROVIO) approach is used (Nguyen et al., 2020). On the other hand, thermal vision can see through the majority of these visual impairments, but is often characterized by low contrast pictures. The determination of thermal-inertial odometry makes use of the modified ROVIO approach. LiDAR Odometry and Mapping can function effectively in most situations where geometrically rich structure is present in the environment, which is often the case in building projects. Performance only degrades when dealing with self-similar geometry, obscurants, and other difficulties. As described, RCSS combines the separate posture updates in a loosely connected manner, resulting in a more robust multi-modal approach.

Planning for the Exploration and Structural Inspection Path

Repeated inspection and data processing are necessary for construction project monitoring. Comprehensive 3D mapping, visual surveillance, coverage with task-specific sensing, and inventory management are particularly crucial. RCSS uses two forms of path planning to carry out these tasks automatically: (1) The robot relies on the Graph-based Exploration Path Planner (GBPlanner) for its inaugural task, specifically designed for scenarios lacking pre-existing models. The approach employs a split architecture, initially creating an exploration path by extending a random network within a small volume around the robot. It evaluates the volumetric gain of these vertices. Only in cases where such a path cannot be established the system switch to an alternative approach. (2) Subsequently, it identifies a previously explored exploration space frontier using the global graph it gradually builds, proceeding to the designated location to execute its assigned task.

The Structural Inspection Planner (SIP) is used for the initial job when a previous model is present, either through blueprints/CAD or a previous exploratory trip. SIP uses a mesh model of the structure and a two-step iterative technique to determine full coverage pathways with the least amount of time spent travelling while taking into account vehicle and sensor restrictions (Shang et al., 2020). By resolving a 3D Art Gallery Problem (AGP), it samples a collection of vertices that serve as perspectives in the first stage of its two-step paradigm to assure complete coverage of the building. The Travelling Salesman Problem (TSP) linked with it is then solved to arrive at an ideal complete coverage path. Notably, given the environment's map, RRT is used to discover that the edges between the AGP-sampled vertices are collision-free.

Experimental Verification

Three field studies were carried out to assess the functionality and potential for use of RCSS. These in particular included

- Autonomous exploration and mapping of the first floor of the University of Nevada, Reno's Arts Building's corridors;
- Autonomous inventory reporting of objects of interest inside another building; and
- Mapping of a construction site related to a community housing project.

The first habitat has its underlying timber-based construction revealed, presenting more intricate geometry with thinner details than the prior two environments' finished constructions. When new activities are to begin in an existing building, a map of the present geometric shape must be gathered. The findings of the autonomous investigation within the university's Arts Building's corridors.

The second experimental research comprised automated inventory management through the use of the robot's online mapping capabilities in conjunction with a Yolov3-based trained item detector for things relevant to building operations (da Silva et al., 2021). As it moves around a building, the robot finds interesting things and automatically records their (a) class, (b) position, and (c) timestamp. The object position is determined by ray casting on the volumetric map representation using the environment's online reconstructed map as a starting point.

Our final two results are based on simulated data. The first deals with the creation of an inspection route based on the relevant results of a prior model of a bridge structure and the capability for systematic inspection. The planner stated in Section III's open-source version was altered to provide consistent coverage in terms of the distance to each mesh facet. The second involves aligning point cloud data collected from various missions and settings in order to spot changes.

This aerial robot is still in the early stages of development, but it relies on a wealth of earlier work in multi-modal localization and mapping as well as autonomous exploration and inspection path planning. Future research will focus on using these resilient and strong algorithms and subsystems with dedicated object detection and location estimation for inventory tracking, as well as the ability to deliver multi-modal maps of construction facilities fully autonomously, quickly, and unrestricted by terrain limitations (Nguyen et al., 2020).

Robotics in Construction: Demolition and Excavation

Indeed, technology has made life easier for many, not just in one, but in almost all, aspects. Global urbanization is still expanding, manifesting itself in new construction, altered infrastructure, and renovated older cities. Much of this advancement is owed to equipment. The workforce and amount of human labor are considerably lowered by the use of machinery. Machines save time and money while also being less expensive.

Uses of Demolition Robot

In the construction sector, demolition robots are frequently utilized for demolition and evacuation. They are essential for transporting bulky items up to higher stories from the lower floor. Ninety percent of the market for robotic construction tools is now occupied by demolition robots (Adami et al., 2021). The popularity of these devices is constantly rising. And when their applications develop further, they may possibly take up greater market share.

Demolition Robot Benefits

Robotic demolition is revolutionizing the building sector. But why are they so useful?

Demolition robots provide several additional benefits due to their heavy-duty nature. Power is one of the most obvious. The robot is significantly smaller and more compact than its competitors, but it has far more power. These tools can work continuously without stopping or pausing. They can perform much better than five- to six-times-larger excavators.

There are several different types of demolition robots. Robots that can be operated remotely are under the category of demolition robots. The robot's ability to be commanded remotely gives its features further control. They are safer than manually operated alternatives and more efficient. The hazards are reduced since operators may manage them from outside the machine. They are protected from dangerous factors and unanticipated collapses.

The majority of demolition robots have a three-armed design. On construction sites, some locations are close together and are difficult to reach. The robot's reach is increased by the three-armed arrangement. These robots have a 360-degree field of view and can access everything. Areas that were previously inaccessible are now easily accessible. The three arms also offer more accuracy and precision for error-free work.

It requires a lot of electricity to evacuate people from buildings and demolish them. The heat produced by this power is intense. The cylinders within heat up as this energy is produced by the construction equipment. Both the engine and the operator are in risk from this. Many demolition robots can withstand heat. The operator, the environment, and the cylinders are all safer this way. They prevent overheating and maintain the safety of every component. The demolition robot lives longer as a result of this.

Robotics in Construction: Building and Assembly

The incorporation of robots technology has significantly changed the construction sector in recent years, which is frequently characterized by labor-intensive operations and specific building requirements. Robotics in construction, especially in the building and assembly sectors, represents a paradigm shift that has the potential to completely transform how structures are put together and built, as illustrated in Figure 6.2. This novel method makes use of automation and robotic technology to improve accuracy, precision, and safety across multiple building processes (Hossain et al., 2020).

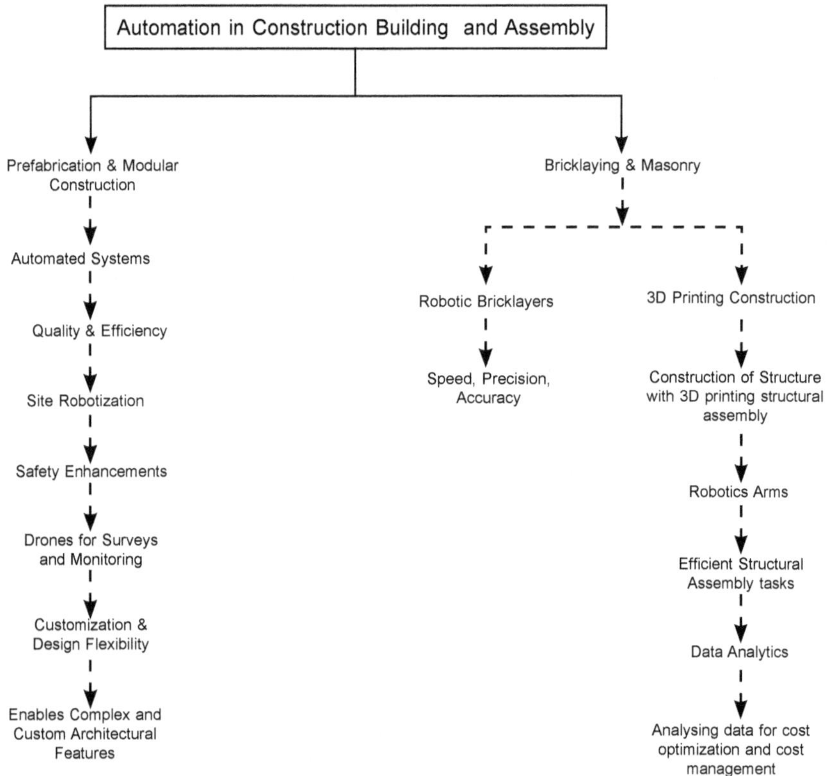

Figure 6.2: The robotics revolution in construction.

Numerous building and assembly-related uses of robotics in construction are changing conventional construction practices in the following ways:

Prefabrication and Modular Construction

Robotics is essential to the prefabrication of building materials including wall panels, roof trusses, and floor systems. Automated systems can accurately produce these components in regulated settings, assuring excellent quality and reducing the time required for on-site assembly (Qureshi et al., 2021).

Bricklaying and Masonry

Robotic bricklayers and masonry systems have become well-known for their exceptional speed and precision when laying bricks. These robots are able to build walls with complex designs while maintaining exacting quality requirements.

3D Printing Construction

3D printing robots have made it possible to construct entire structures subcaste by subcaste. They can use a variety of accoutrements, including concrete and mixes, to produce structurally sound walls, columns, and other structure factors.

Structural Assembly

Robotic arms equipped with advanced detectors and end-effectors are used for precise structural assembly tasks. They can lift and place heavy structure factors with perfection, reducing the threat of mortal error and enhancing safety.

Site Robotization

Robots are employed in colorful construction site tasks, including excavation, material running, and point cleaning. These automated systems enhance productivity by efficiently managing point logistics.

Safety Enhancements

Robotics also contribute to perfecting safety on construction spots. Drones can survey dangerous areas, and robotic systems can handle dangerous tasks, reducing the threat of accidents and injuries.

Customization and Design Flexibility

Robotics technology enables lesser design inflexibility, allowing for the construction of complex and customized architectural features that were preliminarily grueling to execute manually.

Data Analytics

Robotics in construction generates a wealth of data, which can be anatomized to optimize construction processes, manage coffers efficiently, and make data-driven opinions.

Robotics in Construction: Inspection and Maintenance

Numerous ways, including automated brickwork, steel welding, concrete distribution, positioning steel reinforcement, concrete finishing, tile placement, fireproof coating, painting, earthmoving, material handling, and road maintenance, have seen robots used in construction and infrastructure projects (Adepoju, 2022). Robots using a variety of sensors and modes of mobility have been utilized to investigate the built environment. Unmanned aerial vehicles (UAV), unmanned ground vehicles (UGV), maritime vehicles, wall-climbing robots, and robots that can traverse cables are a few examples. A safer alternative to manual examination is robotic inspection. Robotic automation increases inspection frequency and lessens subjectivity in error detection (Shahzad et al., 2021).

Unmanned Aerial Vehicles (UAVs)

UAVs are by far the most often utilized form of robot for surveying and keeping track on the built environment. Association for Unmanned Vehicle Systems International (AUVSI) estimates that the UAV industry is currently worth $11.3 billion in the United States alone and will reach $140 billion in the following ten years (Pathak, 2021). The DJI Phantom 3, DJI Phantom 4, Parrot AR.Drone 2.0, DJI M600 Pro, DJI Matrice 100, FlyTop FlyNovex, DJI Mavic Mini, DJI Mavic Pro, and Tarot FY680 are a few examples of commercial UAVs utilized in research. UAVs, usually referred to as drones, were first utilized in military activities, but they are now increasingly being used in the development and upkeep of civil infrastructures for inspection and monitoring. A wide number of industries, including construction, find them desirable due to their adaptability and cheap operating and maintenance expenses. UAVs are the preferred method of data collecting because to their maneuverability and greater measuring angles (Simmons et al., 2019). Additionally, they are lightweight, and setting them up can be done quickly. Where humans find it difficult to access, they can. UAVs can conduct inspections from heights while keeping people safe from fall dangers. UAVs can investigate objects faster than people can. After severe storms, glass-fronted high-rise structures are frequently damaged. Before the building may be reoccupied, these facades must be assessed. UAVs can deliver reliable information much faster and more frequently than people because of their speed. UAVs can thereby lower the expense and danger associated with inspecting the built environment.

UAVs may be divided into two groups: rotary-wing and fixed-wing UAVs. Despite being speedier, fixed-wing UAVs are unable to hover or take off vertically. Due to its ability to take off vertically from any position, rotary-wing UAV does not require a horizontal plane on which to accelerate (Gu et al., 2017). The quadrotor, which has four rotors, is the most common type of rotary-wing UAV. They are quick and can hover. However, they are slower and have a smaller range. Rotating-wing UAVs are therefore more appropriate for applications related to buildings. However, large, linear infrastructure projects like highways and railroads can benefit from the utilization of fixed-wing UAVs. A lighter-than-air platform is a type of UAV that is relatively uncommon; examples include kites and balloons. These systems have relatively little wind resistance and are much slower.

Unmanned Ground Vehicles (UGVs)

They are also referred to as rovers, UGVs have the most straightforward designs of all the robots covered in this article. These robots function well on flat surfaces but struggle on crowded ones. UGVs can be crawler-mounted or tracked systems, or they can be wheel-driven with pneumatic wheels. Crawlers, which are also seen in military tanks and other large pieces of machinery, provide more grip on slick or muddy ground. Wheeled UGVs are the most effective in terms of power consumption, cost, control, robustness, and speed if the surface conditions are acceptable (Chen et al., 2023). UGVs can be utilized as human assistants for extended inspection cycles due to their reduced power consumption and longer runtime. The Clearpath Jackal and Clearpath Husky are two examples of UGVs that are readily accessible commercially and are utilized for research.

UGVs are the most stable and capable of transporting heavy payloads because of their low center of gravity. Regular fixed cameras and stereo cameras for visual data collection; LiDAR and laser scanners for 3D data capture; mechanical arms for obstacle removal; ground-penetrating RADAR (GPR), ultrasonic sensors, and infrared sensors for behind-wall and under-ground sensing; a graphics processing unit (GPU) for processing; and IMU, GP, and other payloads are some examples of common payloads attached to a UGV.

Due to its simple design and operation, UGVs have been employed in a number of settings besides buildings. They have been used to check storage tanks, HVAC ducts, sewer and water pipes, and even bridge decks. They can also undertake wall building, wall painting, and floor cleaning due to their versatility. Due to their simplicity of use, they provide great test beds for algorithms and sensor design.

UGVs' short height is also a drawback since it restricts their ability to manoeuver in vast corridors or rooms with high ceilings. UGVs are therefore frequently employed in close coordination with other robots, including UAVs.

Robots that Climb Walls

Robots that can scale walls are employed to assess building facades, windows, and outside pipelines. Human inspectors supported by a temporary frame dangling from the structure's top manually assess the outside utilities of high-rise buildings, putting their own safety at great risk. A UAV may be the best instrument for these tasks because of its adaptability and simple setup. However, UAVs are vulnerable to strong winds and governmental regulations. For the examination of outside pipes, Liu created a wall-climbing robot that is hung by a cable. Such robots can transport greater sensor payloads and are significantly safer.

Grip climbing, which uses mechanical actuators and springs to provide gripping motion, is one of the other techniques for climbing walls. These robots can climb the vertical structure without any pre-installed infrastructure. However, they cannot be used on flat surfaces since they require surface protrusions to grasp on to when climbing.

Electromagnets that are controlled by circuits are used by magnetic climbing robots to climb steel buildings. Compared to UAVs, these robots' magnets use extremely little electricity to function, and they can maintain a position forever. Magnetic tracks were used to create a robot with a similar design. The robot was created to examine massive steel storage tanks' interiors. Dust or unevenness on the surface might cause magnetic climbers to lose traction, which is dangerous.

Robots that Climb Cables

Inspecting stay cables in suspension and cable-stayed bridges is a specialized use for cable-crawling robots. These robots are distinct from the wall-climbing robots that are cable-suspended. By using driving rollers, these robots may crawl over the existing steel cables in bridges without the need for additional infrastructure. Despite the fact that drones may be used to securely examine these bridge cables from great heights, they must keep a minimum safe distance to prevent collision. It is difficult to see tiny flaws on the wire surface in images taken from a distance. Additionally, instead of merely inspecting one side at a time, cable-crawling robots outfitted with several cameras, as seen in, may examine the whole surface of the steel cables.

In Japan, a similar cable-crawling robot was also created to examine suspension bridge steel cables. A tethered camera module was dropped from the base platform of their robot to investigate the hanging ropes, and a base platform traversed the major cables of the bridge linking the support tower. The camera module hanging from the base platform performed the actual inspection while the base platform allowed for horizontal movement over the bridge.

Marine Robot

Marine infrastructure including bridge piers, dam embankments, underwater pipes, and offshore constructions are all inspected by marine robots. Unmanned underwater vehicles (UUVs) or autonomous underwater vehicles (AUVs), also known as submersible robots, and unmanned surface vehicles (USVs), sometimes known as surface water platforms (SWPs), are the two main types of marine robots that can be found in the literature (Halder & Afsari, 2023). Unmanned marine vehicles, which can include both USV and UUV, is a phrase that is occasionally used interchangeably.

Microbots with Hinges

Microbots with hinges have many small body segments that are hinged together. These machines mimic the motions of worms or snakes. They don't have wheels; instead, their entire body is equipped with many motorized joints. They may move by either sidewinding like snakes or by contracting and expanding their body like worms. These robots have previously been studied using a variety of propagation techniques, including piezoelectric, hydraulic, pneumatic, and electrical micro-actuators. The best speed and power may be found using electrical micro-motors.

Legged Robot

Compared to other robot types utilized for building inspection and monitoring, legged robots are rather more recent. They move by employing mechanical limbs with many motors in each. Robots with legs can have two (bipedal), four (quadrupedal), or even six (hexapodal) legs. They are highly suited for use on building sites because of their versatility and mobility in a variety of terrains. NASA, MIT, IIT, ETH, Boston Dynamics, Ghost Robotics, ANYbotics, and Unitree have all created several legless robots. However, few have been applied outside of a laboratory context due to their great technical complexity. The literature on commercial legged robots in the construction industry is particularly lacking as a result of their limited market availability. Legged robots have a significant benefit over UGVs in that they can navigate stairs, making multi-story inspections possible.

Combined Robots

Robots of the hybrid class move using a variety of locomotion. These unique robots were created to solve certain issues and are not often offered for sale in the market. According to Morton & Papanikolopoulos (2017), a basic hybrid robot is a wheeled robot with extra rotors on top. In such a robot, the rotors help in inspection at height, obstacle avoidance, and floor change while the wheels give stability on a flat area indoors.

Multiple-Robot Systems

Teams of many robots of the same or other sorts have also been utilized by researchers to investigate and monitor the built environment. Two quadruped robots made up the robotic system, and they cooperated in a 'so-called' master-slave arrangement. The master robot carried a high-performance processing unit for these uses and was in charge of mission planning, task distribution, and the compilation of complete reports. Other payloads that were carried by the slave or secondary robot were a long-range Lidar, a robotic arm, a thermal camera, and a tiny computer for performing local control algorithms. Without the need for a big robotic platform, such devices may transport heavier payloads. The creation of multi-robot collaboration techniques is the main goal of research involving numerous robots.

References

Adami, P., Rodrigues, P.B., Woods, P.J., Becerik-Gerber, B., Soibelman, L., Copur-Gencturk, Y. and Lucas, G. (2021). Effectiveness of VR-based training on improving construction workers' knowledge, skills, and safety behavior in robotic teleoperation. *Advanced Engineering Informatics*, 50, 101431. https://doi.org/https://doi.org/10.1016/j.aei.2021.101431.

Adepoju, O. (2022). Robotic construction technology. *In:* O. Adepoju, C. Aigbavboa, N. Nwulu and M. Onyia (Eds), *Re-skilling Human Resources for Construction 4.0: Implications for Industry, Academia and Government*, pp. 141–169. Springer International Publishing. https://doi.org/10.1007/978-3-030-85973-2_7.

Ammad, S., Alaloul, W.S., Saad, S., Qureshi, A.H., Sheikh, N., Ali, M. and Muhammad Altaf. (2020). Personal protective equipment in construction, accidents involved in construction infrastructure projects. *Solid State Technology*, 63(6). https://solidstatetechnology.us/index.php/JSST/article/view/3766.

Ammad, S., Saad, S., Bashir, M.T., Qureshi, A.H., Altaf, M. and Rasheed, K. (2021). Construction Accidents via Integrating Building Information Modeling (BIM) with Emerging Digital Technologies: A Review. *2021 3rd International Sustainability and Resilience Conference: Climate Change*. https://doi.org/10.1109/IEEECONF53624.2021.9668066.

Babbar, A., Rai, A. and Sharma, A. (2021). Latest trend in building construction: Three-dimensional printing. *Journal of Physics: Conference Series*, 1950(1), 12007. https://doi.org/10.1088/1742-6596/1950/1/012007.

Balaguer, C., Abderrahim, M., Navarro, J.M., Boudjabeur, S., Aromaa, P., Kahkonen, K., Slavenburg, S., Seward, D., Bock, T., Wing, R. and Atkin, B. (2002). FutureHome: An integrated construction automation approach. *IEEE Robotics & Automation Magazine*, 9(1), 55–66. https://doi.org/10.1109/100.993155.

Bock, T. (2007). Construction robotics. *Autonomous Robots*, 22(3), 201–209. https://doi.org/10.1007/S10514-006-9008-5/METRICS.

Chen, G., Jiang, Y., Tang, Y. and Xu, X. (2023). Pitch stability control of variable wheelbase 6WID unmanned ground vehicle considering tire slip energy loss and energy-saving suspension control. *Energy*, 264, 126262. https://doi.org/https://doi.org/10.1016/j.energy.2022.126262.

da Silva, D.Q., dos Santos, F.N., Sousa, A.J., Filipe, V. and Boaventura-Cunha, J. (2021). Unimodal and multimodal perception for forest management: Review and dataset. *Computation*, 9(12). https://doi.org/10.3390/computation9120127.

Davila Delgado, J.M., Oyedele, L., Ajayi, A., Akanbi, L., Akinade, O., Bilal, M. and Owolabi, H. (2019). Robotics and automated systems in construction: Understanding industry-specific challenges for adoption. *Journal of Building Engineering*, 26, 100868. https://doi.org/https://doi.org/10.1016/j.jobe.2019.100868.

Gu, H., Lyu, X., Li, Z., Shen, S. and Zhang, F. (2017). Development and Experimental Verification of a Hybrid Vertical Take-off and Landing (VTOL) Unmanned Aerial Vehicle (UAV). *2017 International Conference on Unmanned Aircraft Systems (ICUAS)*, pp. 160–169. https://doi.org/10.1109/ICUAS.2017.7991420.

Hägele, M., Nilsson, K., Pires, J.N. and Bischoff, R. (2016). Industrial Robotics. *In:* B. Siciliano and O. Khatib (Eds.), *Springer Handbook of Robotics*, pp. 1385–1422. Springer International Publishing. https://doi.org/10.1007/978-3-319-32552-1_54.

Halder, S. and Afsari, K. (2023). Robots in inspection and monitoring of buildings and infrastructure: A systematic review. *Applied Sciences*, 13(4). https://doi.org/10.3390/app13042304.

Hersh, M. (2015). Overcoming barriers and increasing independence: Service robots for elderly and disabled people. *International Journal of Advanced Robotic Systems*, 12(8), 114. https://doi.org/10.5772/59230.

Hossain, M.A., Zhumabekova, A., Paul, S.C. and Kim, J.R. (2020). A review of 3D printing in construction and its impact on the labor market. *Sustainability*, 12(20), 8492. https://doi.org/10.3390/su12208492.

Hussein, M., Nouacer, R., Corradi, F., Ouhammou, Y., Villar, E., Tieri, C. and Castiñeira, R. (2021). Key technologies for safe and autonomous drones. *Microprocessors and Microsystems*, 87, 104348. https://doi.org/https://doi.org/10.1016/j.micpro.2021.104348.

Jo, B.-W., Lee, Y.-S., Khan, R.M., Kim, J.-H. and Kim, D.-K. (2019). Robust construction safety system (RCSS) for collision accidents prevention on construction sites. *Sensors*, 19 (4). https://doi.org/10.3390/s19040932.

K., C.Y., Kinam, K., Shaojun, M. and Jun, U. (2018). A robotic wearable exoskeleton for construction worker's safety and health. *In: Construction Research Congress 2018*, pp. 19–28. https://doi.org/doi:10.1061/9780784481288.003.

Lee, S. and Moon, J.II. (2015). *Case Studies on Glazing Robot Technology on Construction Sites*. IAARC. https://doi.org/10.22260/ISARC2015/0077.

Li, Y. and Liu, C. (2019). Applications of multirotor drone technologies in construction management. *International Journal of Construction Management*, 19(5), 401–412. https://doi.org/10.1080/15623599.2018.1452101.

Madonna, F., Rosoldi, M., Lolli, S., Amato, F., Vande Hey, J., Dhillon, R., Zheng, Y., Brettle, M. and Pappalardo, G. (2018). Intercomparison of aerosol measurements performed with multi-wavelength Raman lidars, automatic lidars, and ceilometers in the framework of INTERACT-II campaign. *Atmospheric Measurement Techniques*, 11(4), 2459–2475. https://doi.org/10.5194/amt-11-2459-2018.

Manuel Davila Delgado, J. and Oyedele, L. (2022). Robotics in construction: A critical review of the reinforcement learning and imitation learning paradigms. *Advanced Engineering Informatics*, 54, 101787. https://doi.org/https://doi.org/10.1016/j.aei.2022.101787.

Mathiassen, K., Fjellin, J.E., Glette, K., Hol, P.K. and Elle, O.J. (2016). An ultrasound robotic system using the commercial robot UR5. *Frontiers Robotics AI*, 3(Feb.), 1–16. https://doi.org/10.3389/frobt.2016.00001.

Melenbrink, N., Werfel, J. and Menges, A. (2020). On-site autonomous construction robots: Towards unsupervised building. *Automation in Construction*, 119, 103312. https://doi.org/https://doi.org/10.1016/j.autcon.2020.103312.

Miozzo, M. and Dewick, P. (2002). Building competitive advantage: Innovation and corporate governance in European construction. *Research Policy*, 31(6), 989–1008. https://doi.org/https://doi.org/10.1016/S0048-7333(01)00173-1.

Morton, S. and Papanikolopoulos, N. (2017). A small hybrid ground-air vehicle concept. *2017 IEEE/RSJ International Conference on Intelligent Robots and Systems (IROS)*, 5149–5154. https://doi.org/10.1109/IROS.2017.8206402.

Naeem, U.H., Bilal, M., Syed, F., Rasheed, K. and Saad, S. (2022). A Multilayer Encryption Model to Protect Healthcare Data in Cloud Environment. *2022 International Conference on Data Analytics for Business and Industry, ICDABI 2022*. https://doi.org/10.1109/ICDABI56818.2022.10041708.

Nguyen, H., Mascarich, F., Dang, T. and Alexis, K. (2020). *Autonomous Aerial Robotic Surveying and Mapping with Application to Construction Operations*. https://doi.org/10.48550/arXiv.2005.04335.

Pathak, R. (2021). *Development of Data Processing Tool for Precision Agriculture and Delivery System to End User*. Ph.D. project for Mississippi State University.

Qiu, S., Pei, Z., Wang, C. and Tang, Z. (2023). Systematic review on wearable lower extremity robotic exoskeletons for assisted locomotion. *Journal of Bionic Engineering*, 20(2), 436–469. https://doi.org/10.1007/s42235-022-00289-8.

Qureshi, A.H., Alaloul, W.S., Manzoor, B., Saad, S., Alawag, A.M. and Alzubi, K.M. (2021). Implementation Challenges of Automated Construction Progress Monitoring under Industry 4.0 Framework towards Sustainable Construction. *2021 Third International Sustainability and Resilience Conference: Climate Change*, pp. 322–326. https://doi.org/10.1109/IEEECONF53624.2021.9668074.

Qureshi, A.H., Alaloul, W.S., Murtiyoso, A., Hussain, S.J., Saad, S. and Oad, V.K. (2022). Evaluation of 3D Model of REBAR for Quantitative Parameters. *The International Archives of the Photogrammetry, Remote Sensing and Spatial Information Sciences*, XLVIII-2/W, pp. 215–220. https://doi.org/10.5194/isprs-archives-XLVIII-2-W1-2022-215-2022.

Qureshi, A.H., Alaloul, W.S., Wing, W.K., Saad, S., Alzubi, K.M. and Musarat, M.A. (2022). Factors affecting the implementation of automated progress monitoring of rebar using vision-based technologies. *Construction Innovation* (ahead-of-print). https://doi.org/10.1108/CI-04-2022-0076.

Saad, S., Alaloul, W.S., Ammad, S., Qureshi, A.H., Altaf, M. and Rasheed, K. (2020). Design Phase Carbon Emission Prediction Using a Visual Programming Technique. *2020 2nd International Sustainability and Resilience Conference: Technology and Innovation in Building Designs*. https://doi.org/10.1109/IEEECONF51154.2020.9319978.

Saad, S., Alaloul, W.S., Ammad, S., Altaf, M. and Qureshi, A.H. (2021). Identification of critical success factors for the adoption of Industrialized Building System (IBS) in Malaysian construction industry. *Ain Shams Engineering Journal*, 13(2), 101547. https://doi.org/https://doi.org/10.1016/j.asej.2021.06.031.

Saad, S., Alaloul, W.S., Ammad, S. and Qureshi, A.H. (2022). A qualitative conceptual framework to tackle skill shortages in offsite construction industry: A scientometric approach. *Engineering, Construction and Architectural Management*, 29(10), 3917–3947. https://doi.org/10.1108/ECAM-04-2021-0287

Shahzad, L., Rasheed, K., Ahmad, I., Haider, S.W., Saad, S. and Ahmed, E. (2021). Cost Effective Accident Prevention System for Vehicles. *2021 International Conference on Decision Aid Sciences and Application, DASA 2021*. https://doi.org/10.1109/DASA53625.2021.9682266.

Shang, Z., Bradley, J. and Shen, Z. (2020). A co-optimal coverage path planning method for aerial scanning of complex structures. *Expert Systems with Applications*, 158, 113535. https://doi.org/https://doi.org/10.1016/j.eswa.2020.113535.

Simmons, B.M., McClelland, H.G. and Woolsey, C.A. (2019). Nonlinear model identification methodology for small, fixed-wing, unmanned aircraft. *Journal of Aircraft*, 56(3), 1056–1067. https://doi.org/10.2514/1.C035160.

Stone, W.L. (2018). *The History of Robotics*. CRC Press Boca Raton, FL.

Thompson, I. (2014). *Landscape Architecture: A Very Short Introduction* (Vol. 387). Oxford University Press, USA.

Valero, E., Sivanathan, A., Bosché, F. and Abdel-Wahab, M. (2016). Musculoskeletal disorders in construction: A review and a novel system for activity tracking with body area network. *Applied Ergonomics*, 54, 120–130. https://doi.org/https://doi.org/10.1016/j.apergo.2015.11.020.

CHAPTER
7

AI-assisted Building Design

Syed Saad, Muhammad Haris, Syed Ammad and Kumeel Rasheed

Introduction

AI-assisted Building Design is transforming the field of architecture, revolutionizing how architects conceive and create structures. Integrating Artificial Intelligence (AI) technologies in the design process has opened up exciting opportunities for architects to explore innovative and eco-friendly design solutions. Construction industry professionals admire the evolution of digital data-acquisition technologies in monitoring processes due to efficient and efficacious outcomes (Qureshi et al., 2020). The remarkable progress in AI has empowered architects to harness computational power and data-driven algorithms for generating and assessing design options. To inform the design process and improve building performance, AI can evaluate enormous volumes of data, including building performance, environmental conditions, and user preferences. Using AI, architects may create structures that are not only visually appealing but also ecologically friendly, energy-efficient, and suited to specific user needs.

Managing a construction project to satisfy the requirements of a diverse base of stakeholders during these times of rapid change is an extremely challenging objective (Syed et al., 2022). AI-assisted building design offers a significant advantage in amplifying architects' creative capabilities, revolutionizing the traditional design process. Leveraging sophisticated algorithms, AI systems possess the capacity to generate diverse design alternatives guided by predefined parameters and constraints. This feature empowers architects to embark on a journey of exploration through an extensive array of design possibilities, expanding the horizons of their creativity. The integration of AI in the design workflow encourages a collaborative and symbiotic relationship between human designers and AI technology. In this context, AI operates as a creative partner, augmenting the architect's expertise and vision by offering novel insights, innovative ideas, and design options that may not have been previously contemplated.

AI algorithms can uncover novel and cutting-edge ideas, pushing the boundaries of design inquiry, by evaluating massive amounts of data and drawing from a comprehensive database of architectural patterns and historical precedents. Architects can also swiftly iterate and enhance their designs due to the iterative nature of AI algorithms, which makes the design process more effective and dynamic.

This cooperative relationship between AI technology and architects improves the quality of design solutions and makes it easier to make wise decisions. AI-generated design alternatives serve as a valuable resource for architects to critically evaluate and compare various proposals against project requirements and objectives. Such a data-driven approach empowers architects to make well-informed choices, fostering the development of optimized and contextually relevant designs. Moreover, the integration of AI in building design alleviates the burden of repetitive and time-consuming tasks, liberating architects to concentrate on the more creative and strategic aspects of their work. By delegating routine operations to AI systems, architects can redirect their efforts towards conceptualizing innovative solutions, refining designs, and engaging in thoughtful design critique (Rigillo, 2023).

While AI-driven design presents unprecedented opportunities, it is essential to recognize that the successful implementation of AI in architecture relies on a human-centric approach. Architects remain the primary decision-makers, exercising their judgment and intuition to refine AI-generated options and infuse the designs with their unique perspectives. Additionally, ethical considerations surrounding AI implementation must be diligently addressed, ensuring transparency, fairness, and accountability in the design process. Furthermore, AI-driven building design significantly enhances the design process's efficiency. AI algorithms can automate repetitive and time-consuming tasks, such as generating floor plans, optimizing building layouts, and conducting energy simulations (Baduge et al., 2022). This automation liberates architects' time, enabling them to concentrate on higher-level design decisions and problem-solving. Additionally, AI can optimize building systems, such as heating, ventilation and air conditioning (HVAC) and lighting, to maximize energy efficiency and occupant comfort (Rigillo, 2023).

However, the integration of AI in building design raises essential ethical and societal concerns. Relying solely on AI algorithms prompts questions about the role of human creativity and the potential for biased design outcomes. Striking a balance between AI capabilities and human architects' expertise and intuition is crucial to ensure that the design process remains human-centered and reflective of societal values (Rigillo, 2023).

AI-assisted building design offers a significant advantage by enhancing architects' creative capacities. Through AI algorithms, design alternatives are generated based on predefined parameters and constraints, enabling architects to explore a broad spectrum of possibilities. This collaborative approach between human designers and AI systems fosters a symbiotic relationship, where AI serves as a creative partner, providing inspiration and generating design options that may have been overlooked otherwise (Bölek et al., 2023).

Furthermore, AI-assisted building design leads to a notable improvement in design process efficiency. Repetitive and time-consuming tasks, such as generating floor plans, optimizing building layouts, and conducting energy simulations, can be automated using AI algorithms. This automation liberates architects' time, enabling them to focus on higher-level design decisions and problem-solving. Additionally, AI lends a helping hand in optimizing building systems like HVAC and lighting, effectively maximizing energy efficiency and occupant comfort (Bölek et al., 2023). The potential for AI to completely transform the architectural industry is enormous. Architects may increase their capacity for creativity, improve design effectiveness, and produce user-centered, sustainable structures by utilizing AI technologies. However, it is vital to carefully consider the ethical and societal implications of AI integration to ensure that the design process remains human-centered.

In conclusion, AI-assisted building design presents immense potential for revolutionizing the architecture field. By leveraging AI technologies, architects can enrich their creative capacities, enhance design efficiency, and develop sustainable buildings that cater to user needs. However, it is vital to carefully consider the ethical and societal implications of AI integration to ensure that the design process remains human-centered and aligned with societal values.

AI-based Building Design Techniques

The advent of AI-driven building design processes has ushered in a transformative era for the field of architecture, bestowing architects with innovative tools and methodologies to conceive creative and sustainable solutions. The usage of new technology has drastically altered the way, work is handled and performed out in recent times (Ammad, Alaloul et al., 2021). Through the seamless integration of AI technology, architects now possess the capability to optimize building performance, ignite innovation, and amplify productivity, revolutionizing the way architectural projects are conceived and realized.

AI's integration in building design encompasses a diverse array of techniques and approaches, each contributing to a more holistic and efficient design process (Dwivedi et al., 2021). Among these, generative design emerges as a prominent and powerful method, where AI algorithms take center stage in generating a multitude of design options based on predefined parameters and constraints. This revolutionary process empowers architects to venture into an expansive realm of design possibilities, unearthing optimized solutions that might have remained undiscovered through traditional approaches.

Generative design, in essence, serves as a catalyst for architectural creativity, extending the boundaries of conventional design methodologies. By leveraging AI's immense computational capabilities, architects can swiftly generate and evaluate a vast range of design alternatives, thereby expediting the exploration of innovative concepts. This iterative process enables architects to break free from the constraints of time-consuming manual design iterations, paving the way for efficient and iterative ideation.

Moreover, the implementation of generative design fosters an interactive and collaborative dialogue between architects and AI technology. Rather than replacing human creativity, AI serves as a creative partner, augmenting architects' expertise and intuition with its data-driven insights.

As architects input their design goals, preferences, and constraints into the AI algorithms, the resulting design alternatives are curated to align with the architects' vision, ensuring a seamless integration of human design sensibilities and AI-generated solutions.

Beyond its creative potential, the application of AI in building design yields substantial benefits in terms of building performance and sustainability. AI-driven optimization processes can analyze and identify the most efficient and resource-friendly configurations, ultimately leading to energy-efficient and environmentally conscious architectural designs.

The ability to explore and fine-tune various design permutations empowers architects to deliver buildings that strike an optimal balance between functionality, aesthetics, and sustainability.

However, while the integration of AI in building design offers remarkable opportunities, it is essential to navigate certain challenges and considerations.

As AI algorithms rely heavily on the data they are trained on, ensuring the ethical and unbiased sourcing of data is paramount to prevent potential biases in the design outcomes.

Furthermore, maintaining transparency in the decision-making process and preserving the architect's agency in guiding the design exploration are vital to uphold the integrity of the creative process.

Performance optimization is a key AI-based method that uses AI algorithms to assess building performance data including energy usage, thermal comfort, and daylighting to improve design choices and maximize building performance. Architects may employ AI to create structures that are environmentally responsible, energy-efficient, and responsive to customer expectations (Mehmood et al., 2019).

AI integration also benefits parametric design, as AI algorithms can generate parametric models to explore different design variations based on changing parameters. This facilitates rapid iteration and evaluation of design options, leading to more efficient and effective solutions.

Furthermore, AI has found applications in building information modeling (BIM), where AI algorithms automate tasks like clash detection, quantity takeoff, and cost estimation, thereby enhancing efficiency and accuracy in the design process. The integration of AI in BIM platforms streamlines workflows and enables architects to make better-informed design decisions.

Incorporating Natural Language Processing (NLP) techniques into AI-based building design has also proved valuable. NLP algorithms analyze textual data, such as design briefs and user requirements, to extract valuable insights and inform the design process (Ding et al., 2022). This helps architects better understand user needs and preferences, leading to more user-centric designs.

1. *Some AI-based Building Design Techniques Generative Design:* AI algorithms can generate multiple design options based on predefined parameters and constraints. These algorithms explore a wide range of possibilities and can assist architects in finding innovative and optimized design solutions.
2. *Performance Optimization:* AI algorithms can analyze building performance data, such as energy consumption, thermal comfort, and daylighting, to optimize building design for energy efficiency and occupant comfort. These algorithms can identify the most effective design strategies and configurations.
3. *Parametric Design:* AI algorithms can be used to create parametric models that allow architects to explore different design variations based on changing parameters. This approach enables architects to quickly iterate and evaluate design options.
4. *Building Information Modeling (BIM):* AI can be integrated into BIM platforms to automate tasks such as clash detection, quantity takeoff, and cost estimation. AI algorithms can analyze data contained within a BIM model to generate insights and aid decision-making during the design process.
5. *Natural Language Processing (NLP):* Textual data, such as design briefs, user requirements, and feedback, can be analyzed using NLP techniques to extract significant insights and inform the design process. NLP algorithms can help architects understand user needs and preferences more effectively.

The integration of AI-based solutions is reshaping the building design process, empowering architects to delve into a multitude of design possibilities as shown is Figure 7.1, enhance building performance, and make well-informed decisions (Parihar et al., 2023). Nevertheless, it is crucial to emphasize that these methodologies methodologies as shown in Table 7.1 must complement human skill and creativity to ensure a design process that remains human-centered and ethically responsible.

Building Information Modeling (BIM) and AI

BIM and AI, two game-changing technologies that have come together, have significantly changed the construction business. The collaboration practices between architects, engineers, and construction professionals have transformed throughout the whole building lifecycle, from design through construction and management. BIM is a strong digital representation of building structures and capabilities. Parallel to this, AI, a field of computer science devoted to the creation of intelligent systems capable of learning and solving problems, has become a powerful force with the capacity to optimize various aspects of construction processes.

The integration of BIM technology has fundamentally reshaped the traditional approach to building design and construction. By creating a comprehensive and detailed digital model of the building, BIM enables seamless communication and collaboration among different stakeholders involved in the project (Ammad,

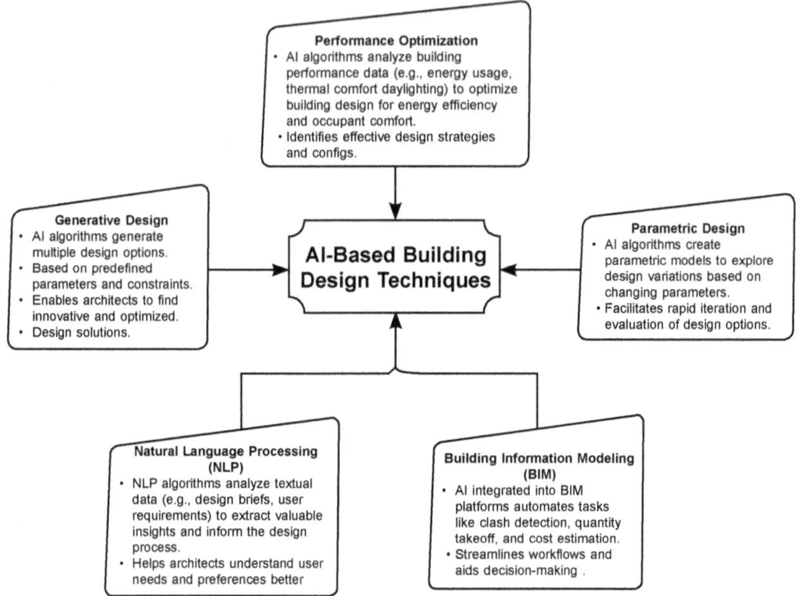

Figure 7.1: AI-based building design

Saad et al., 2021). Architects, engineers, and construction professionals can now work together in a highly coordinated manner, accessing and contributing to a centralized repository of data and information. This collaborative environment fosters efficient decision-making, minimizes conflicts, and enhances the overall project coordination, resulting in streamlined construction processes and reduced project delays (Rasheed et al., 2022).

Moreover, BIM's capabilities extend beyond the design and construction phases to encompass the management and maintenance of buildings throughout their operational lifespan. With BIM-generated data, facility managers can access critical information about the building's components, systems, and performance, facilitating proactive maintenance and maximizing operational efficiency. AI algorithms and machine learning (ML) techniques enable construction professionals to analyze vast datasets and extract valuable insights, accelerating project planning and execution. Predictive analytics powered by AI can anticipate potential project risks, optimize resource allocation, and forecast construction schedules more accurately, mitigating delays and cost overruns.

AI applications in construction span various domains, such as robotics, autonomous vehicles, and predictive maintenance. Robotic systems equipped with AI can perform repetitive and hazardous tasks with precision and efficiency, enhancing worker safety and productivity. Autonomous vehicles and drones are employed for site inspection and data collection, reducing manual efforts and enabling real-time monitoring of construction progress.

Table 7.1: AI-based building design techniques

AI-based building design techniques	Explanation	Comparison
Generative Design	AI algorithms generate multiple design options based on predefined parameters and constraints. This process allows architects to explore a wide range of possibilities and discover innovative and optimized design solutions.	Offers a vast array of design options, fostering creativity and exploration. Optimizes the design process, resulting in efficient ideation. Requires clear definition of parameters and constraints to yield meaningful results.
Performance Optimization	AI algorithms analyze building performance data, such as energy usage, thermal comfort, and daylighting, to optimize building design for energy efficiency and occupant comfort. These algorithms identify effective design strategies and configurations to enhance building performance.	Enhances building sustainability and energy efficiency. Improves occupant comfort and well-being through data-driven design. Requires accurate and comprehensive performance data for reliable optimization.
Parametric Design	AI algorithms create parametric models that enable architects to explore different design variations based on changing parameters. This approach facilitates rapid iteration and evaluation of design options, leading to more efficient and effective solutions.	Facilitates rapid exploration and evaluation of design alternatives. Allows for dynamic adjustments based on parameter changes. Requires defining relevant parameters and their ranges for meaningful outcomes.
Building Information Modeling	AI integration in BIM platforms automates tasks like clash detection, quantity takeoff, and cost estimation. By analyzing data contained within a BIM model, AI algorithms generate insights and aid decision-making during the design process, enhancing efficiency and accuracy.	Improves workflow efficiency and accuracy. Enables better-informed design decisions through data analysis. Requires comprehensive and error-free BIM data for reliable results.
Natural Language Processing	NLP algorithms analyze textual data, such as design briefs and user requirements, to extract valuable insights and inform the design process. This helps architects better understand user needs and preferences, leading to more user-centric designs.	Enhances user-centric design through better understanding of user requirements. Facilitates effective communication and collaboration. Requires high-quality and relevant textual data for meaningful insights.

Furthermore, AI's ability to learn from historical data and adapt to dynamic construction environments enables continuous improvement and optimization in construction processes. AI-driven insights lead to data-driven decision-making, enhancing project outcomes, and contributing to a more sustainable and resilient built environment.

While the convergence of BIM and AI offers unparalleled opportunities for the construction industry, it also brings forth certain challenges. Integrating these sophisticated technologies into existing workflows and practices requires organizational readiness and investments in technology and training. Additionally, ensuring data security and privacy in the context of AI-driven analytics demands robust governance frameworks and ethical considerations.

In conclusion, the fusion of BIM and AI technologies has ushered in a new era of innovation and efficiency in the construction sector. BIM's collaborative capabilities and AI's intelligent automation are reshaping the industry's landscape, enhancing project coordination, optimization, and sustainability. Embracing these transformative technologies and harnessing their potential will empower construction professionals to tackle the challenges of the modern built environment and create structures that are more resilient, sustainable, and responsive to the needs of society.

The amalgamation of BIM and AI presents a promising alliance, unlocking unprecedented possibilities to enhance project outcomes, elevate decision-making processes, and streamline construction workflows. By harnessing the data-rich capabilities of BIM and coupling them with the cognitive prowess of AI, construction projects can be executed with heightened precision, cost-effectiveness, and environmental sustainability.

The collaboration between BIM and AI assumes the role of a revolutionary catalyst as the construction industry sets off on its journey toward digital transformation (Yadav et al., 2023), ushering in a new era of creativity and efficiency throughout the whole project lifecycle. Construction professionals may trace a progressive path towards a more sustainable, resilient, and intelligent built environment by fully understanding and utilizing the promise of these cutting-edge technologies.

The design, engineering, and construction industries have long been very interested in automation and information technology (Hannan Qureshi et al., 2023). Robotics and BIM have been discussed, but their actual application has been constrained by organizational hurdles, cultural barriers, and technical limitations. The industry's adoption of BIM and robotics has been hampered by these issues.

For instance, Eastman's vision of a computerized Building Design System (BDS) resembling today's BIM was proposed as early as 1975 (Sacks et al., 2020). However, it took 25 years for the basic BIM functions to reach the market, and some of the envisioned functionalities, like automated building code checking, are yet to be fully realized. Similarly, implementing robotics in construction has been challenging, with only a few practical and cost-effective solutions emerging after decades of research.

Nevertheless, within the past five years, a notable influx of innovative companies has emerged, specializing in applying information and automation technologies from other industries to construction. Supported by venture capitalists, academic research, and incubator programs, these startups form an ecosystem commonly known as "Construction Tech". The amount of venture capital invested in Construction Tech has grown significantly, indicating increasing interest in this domain.

The newfound practicality of these Construction Tech companies can be attributed not only to the maturation of core technologies but also to the comprehensive building information available in BIM environments. Here, a BIM environment encompasses various BIM tools, platforms, servers, libraries, and workflows within a project or organization. Building information, which encompasses building product designs and construction process plans, serves as the foundation upon which Construction Tech applications build and deliver value. These applications can be broadly categorized into four types: software tools for design and construction management, BIM-to-field tools, robotic applications for on-site construction operations, and field-to-BIM tools.

BIM plays a crucial role as an integrating technology, bridging information gaps within construction projects and supporting (Ammad, Alaloul, et al., 2021). Construction Tech innovations. However, BIM platforms, tools, and processes may require further development to fully support the emerging technology innovations.

The architecture, engineering, and construction (AEC) business has a lot to look forward to with the development of BIM and its integration with AI. Advanced software tools utilizing computer vision, rule-inferencing, ML, case-based reasoning, and other AI techniques have the capacity to intelligently edit BIM models (Saad, Alaloul, Rasheed et al., 2022), which contain extensive building and infrastructure information. The envisaged applications of this synergy between BIM and AI are extensive, ranging from design support and automation to construction planning, safety control, sustainability assessments, and facility management through digital twins (Sacks et al., 2020). Researchers worldwide have long sought to realize these AI-driven tools in the construction domain. Early attempts in the 1980s and 1990s focused on applying expert systems and case-based reasoning to various tasks. However, the limitations of CAD technology, which represented building information graphically and symbolically rather than in an object-oriented manner, hindered progress. The subsequent development of BIM model authoring platforms and the open object-oriented schema for representing buildings and infrastructure (IFC data model) paved the way for commercial implementation of innovations.

Startups have emerged as major drivers of innovation in the Construction Tech ecosystem due to organizational fragmentation within the industry and the need for expertise in adapting technologies from other domains. Fragmentation is evident in vertical, horizontal, and longitudinal aspects, which necessitate new integrated organizations with significant startup costs and risks to foster systemic innovations (Pan & Zhang, 2023).

The "House of Construction Tech" model represents the essential components for success in the sector. At its base lies the BIM environment, encompassing technology, processes, and skilled personnel capable of implementing information management standards and execution plans (Saad, Alaloul, Ammad et al., 2022). Building information, presented in a form suitable for software manipulation, is vital for almost all Construction Tech innovations, underpinning the growth of the sector.

However, certain limitations still exist in BIM technology and processes that restrict the full potential of Construction Tech applications. These limitations include

- Information interoperability challenges
- Difficulties in framing model data for machine learning
- Need for an intelligent digital twin platform to support integration of field-to-BIM tools.

The success of Construction Tech startups rests on incorporating pillars of entrepreneurship, a genuine industry process needs, application of innovative technology, and a viable business model. Embracing the theoretical aspects of design, information and data science, and production in construction is crucial for sustained progress. The House of Construction Tech serves as a checklist and predictor of success for startup companies, highlighting the significance of incorporating robust foundations and pillars. Ignoring essential aspects, such as the conceptualization of production in construction and reliance on 2D drawings instead of BIM models, may hinder long-term success. Successful innovators must identify real business process needs and market demands before developing AI-driven applications to unlock the full potential of BIM and AI in shaping the bright future of Construction Tech (K. Wang et al., 2023).

Case Studies of AI-assisted Building Design

AI has emerged as a powerful catalyst in the field of architecture, effectuating significant changes in design, construction, and urban planning. Numerous case studies exemplify the efficacious integration of AI within architecture, exemplifying its prowess in design optimization, enhanced energy efficiency, and valuable contributions to sustainable development (Lluís Zamora Mestre et al., 2019).

In the realm of building design optimization, AI-driven generative design platforms have gained traction. These platforms employ advanced algorithms to generate multiple design options based on various parameters such as site conditions, user preferences, and environmental considerations. Architects can leverage these generated designs to explore innovative possibilities and select the most efficient and creative solutions for their projects (Lluís Zamora Mestre et al., 2019).

Sustainability and energy efficiency have become key priorities in modern architecture. AI's data analysis capabilities have enabled the development of smart buildings that continuously monitor energy usage, HVAC systems, and lighting. Through real-time data analysis, AI can adjust optimize energy consumption without compromising occupants' comfort, resulting in eco-friendly and cost-effective structures (Lluís Zamora Mestre et al., 2019).

AI has also extended its influence to urban planning and smart cities. By analyzing vast datasets related to traffic flow, population density, and infrastructure requirements, AI algorithms aid in designing efficient transportation systems and optimizing urban layouts. Smart cities harness the power of AI to create sustainable and technologically advanced urban environments that enhance the overall quality of life for their residents (Allam & Dhunny, 2019).

Additionally, AI has been instrumental in historic building preservation and restoration. ML algorithms analyze historical data and architectural patterns to facilitate restoration efforts, preserving cultural heritage and architectural landmarks for future generations.

Furthermore, AI plays a role in construction and project management. Construction robots and drones equipped with AI capabilities can execute tasks such as bricklaying and site inspections with precision and efficiency. AI-powered project management tools assist in predicting project timelines, resource allocation, and risk management, resulting in improved planning and project execution (Lluís Zamora Mestre et al., 2019).

Moreover, the development of natural language interfaces in architectural design has simplified communication between architects and clients. AI-powered systems interpret clients' preferences and requirements expressed in natural language, transforming them into initial design concepts and facilitating a more intuitive and collaborative design process (Lluís Zamora Mestre et al., 2019).

In conclusion, the case studies of AI in architecture illustrate the technology's transformative potential in various aspects of the industry. AI's ability to optimize designs, enhance energy efficiency, contribute to sustainable development, and streamline construction processes has paved the way for a more innovative, eco-friendly, and technologically advanced architectural landscape. As AI continues to evolve, its integration into architecture promises even more profound impacts on how we design and construct buildings and shape the cities of the future.

AI-assisted Building Design: Conceptual and Schematic Design

Considering the critical aspects of managing the conceptual and schematic stages of building design, which represent the foundational phases of the overall design process. These initial stages significantly influence the subsequent detailed design and play a pivotal role in determining the project's overall success and cost (Baldwin et al., 2010).

To address the complexities inherent in the various stages of the design process, a generic conceptual and schematic data flow model is commonly introduced. This model serves as a fundamental framework for monitoring the seamless flow of design information, thereby offering design managers a basis for analysis and decision-making. Through the integration of data from this model with other tools and techniques, a more comprehensive understanding of critical issues, such as missing information, phased information release, uncertainties, and variations in information quality, can be achieved (Baldwin et al., 2010).

The generic conceptual and schematic data flow model provides a structured representation of the flow of design information throughout the design process. It outlines the sequence and interactions of various data inputs, transformations, and outputs at each stage of the design workflow. This allows design managers to gain insights into the progression of design information, identify potential bottlenecks or delays, and make informed decisions to enhance efficiency and productivity (Baldwin et al., 2010).

By incorporating data from other sources and complementary tools, the model becomes enriched with additional context and insights. The data flow model can incorporate data from project management software, BIM software (Building Information Modeling), and other design analysis tools. This integration gives the design process a more comprehensive perspective, enabling a deeper comprehension of the interdependencies across various elements and disciplines.

This method's capacity to identify possible problems with data correctness, completeness, and consistency is one of its main advantages. Design managers can locate locations where information may be partial or missing by visualizing the data flow. This insight enables them to proactively address these gaps, ensuring that all necessary information is available at each stage of the design process (Baldwin et al., 2010).

Furthermore, the data flow model facilitates the evaluation of phased information release, a common practice in complex design projects. It allows design managers to track the sequential release of information as it progresses through various design phases. This monitoring capability is particularly valuable in projects with multiple stakeholders, where information needs to be shared incrementally while maintaining coordination and control (Baldwin et al., 2010).

Uncertainties and variations in information quality are inherent challenges in the design process. The data flow model helps in visualizing these uncertainties and variations, allowing design managers to assess their potential impact on the overall project. Armed with this knowledge, they can implement risk mitigation strategies and allocate resources more effectively to manage potential disruptions.

The introduction of a generic conceptual and schematic data flow model provides an essential tool for design managers to navigate the complexities of the design process. By visualizing the flow of design information and integrating data from various sources, this model enables a comprehensive understanding of critical issues and supports data-driven decision-making. As the construction industry embraces digital transformation and information-driven approaches, leveraging such models becomes increasingly vital for ensuring efficient and successful project outcomes. (Baldwin et al., 2010).

Additionally, two essential tools, the Design Structure Matrix (DSM) and simulation models can be used. The DSM aids in identifying iterative tasks and highlighting interfaces between them, helping design managers identify pivotal points in the design process to avoid inefficiencies and design changes (Baldwin et al., 2010).

A simulation model proves valuable in assessing complex scenarios related to information gaps, assumptions, information release, information quality, gatekeeping, uncertainties, and resource management. However, the simulation's feasibility during the conceptual stage is limited due to its inherent nature, where substantial time is spent on creative thinking and waiting for client decisions, making simulations impractical and offering no significant benefits (Baldwin et al., 2010) .

In conclusion, the effective management of the conceptual and schematic stages of building design is crucial. The generic data flow model, along with the DSM and simulation tools, contributes to improved design management practices and offers valuable insights into information exchange dynamics. Understanding and optimizing these early design phases are vital for ensuring successful architectural projects and minimizing costly redesign work.

AI-assisted Building Design: Detailed Design and Analysis

The architectural design process follows a somewhat linear trajectory, yet it presents numerous opportunities for architects to harness the potential of AI. Many architects have already embraced this cutting-edge technology, incorporating text-to-image generators and AI rendering plugins into their design workflows. The advent of new AI tools such as MidJourney, Dall-E, and ChatGP0054 has expanded the horizons for architects to seamlessly integrate AI into their design processes. Leveraging AI in architectural design has become increasingly accessible and user-friendly. By integrating AI, architectural design efficiency can be significantly enhanced, with AI tools contributing to streamlining mundane tasks like design prototyping, risk management, and cost control. These AI-powered solutions offer promising prospects for simplifying and improving various phases of the architectural design process.

How architects can easily integrate AI into the six phases of the architectural design process is discussed below.

Phases of the Architectural Design Process

1. Pre-design
2. Survey and schematic design
3. Design development

4. Planning application
5. Construction documents
6. Construction administration

How to Integrate AI in Six Phases of the Architectural Design Process

Phase 1: Pre-Design (PD)

The pre-design phase in architecture represents the initial and foundational stage of the architectural design process. At this critical stage, architects and designers embark on defining the project's scope and objectives, while also diligently collecting pertinent information concerning the site and the specific requirements of the building's intended users. Collaboratively, architects, designers, and clients collaborate to establish project goals, identify relevant stakeholders, devise a suitable budget, formulate a comprehensive program (McKenna-Cress & Kamien, 2013), and ascertain and adhere to pertinent regulatory requirements, design establishes the foundation for the entire architectural project, therefore its importance cannot be overstated. All other phases and sections of the project are built on the meticulous planning and data gathering that take place during this phase. Therefore, the effectiveness and thoroughness of the pre-design stage are essential to the overall success of the architectural project.

AI Uses in Pre-Design

AI can play a pivotal role in architectural program analysis, particularly through the application of Natural Language Processing (NLP) algorithms. By leveraging extensive datasets of architectural programs, these algorithms can effectively analyze and interpret the contents of a given program. Through NLP, AI algorithms can discern the specific spatial requirements and functions essential for a building. Additionally, they can identify the interconnections and relationships among these spaces. This valuable information serves as a guide in the design process, empowering architects to ensure that the building fulfills the needs of its users while functioning optimally.

Moreover, AI's analytical capabilities extend to scrutinizing the relationships between various elements within the architectural program, thereby uncovering potential conflicts or discrepancies. For instance, an AI system may detect instances where the size or location of a particular space clashes with the prerequisites of a specific function or activity.

By employing AI in architectural program analysis, designers gain invaluable insights that aid in creating efficient and harmonious building designs (Hafez et al., 2023). The integration of AI technologies in this realm holds the promise of enhancing the architectural design process, yielding structures that are precisely tailored to the users' requirements and seamlessly interwoven with their intended functions.

Phase 2: Schematic Design (SD)

The schematic design phase represents the second stage in the architectural design process. This crucial phase revolves around formulating conceptual ideas for the building's layout, structure, and functionality. During this stage, architects and designers are engaged in generating multiple design options, evaluating existing designs and precedents, refining the concepts, and creating preliminary drawings that illustrate the envisioned appearance and operational aspects of the building.

Client involvement holds paramount importance throughout the schematic design phase, as their input and feedback play a pivotal role in shaping and refining the design. Collaboratively, architects and clients work together to fine-tune the concepts and ensure that the proposed design aligns seamlessly with the client's vision and requirements.

By fostering active client engagement during this phase, architects can better address specific preferences, functional needs, and project objectives. This collaborative approach empowers clients to actively participate in the design development process, leading to a more satisfying and customized architectural solution that meets their unique needs and aspirations. The resulting design reflects a harmonious synthesis of the client's vision and the architect's expertise, laying the groundwork for the subsequent stages of the architectural design process.

AI Uses in Schematic Design

- **Site Assessment:** AI can analyze building sites and gather data on factors like topography, solar exposure, wind patterns, and proximity to amenities. This information helps architects make informed decisions about building design and orientation, optimizing sustainability and functionality. AI's rapid data processing expedites the site analysis phase, ensuring efficient and data-driven decision-making for better architectural outcomes.
- **Schematic Floor Plans:** Integrating text-based AI software like ChatGPT with generative design tools allows architects to interact using natural language commands. This enhances the software's adaptability, enabling architects to communicate preferences and constraints more efficiently. The fusion streamlines the design process, encourages exploration, and accelerates the generation of innovative floor plans.
- **Look and Feel Conversations (Concept Images):** Architects and designers are leveraging AI software like MidJourney and Night Cafe to generate design iterations based on text-based prompts. By incorporating these tools, they can produce detailed concept imagery during the schematic design phase, enabling more effective communication with clients. This imagery forms the foundation for the building's overall aesthetic, aiding in decision-making about materials, lighting, color, and geometry at an early stage of the design process. The use of AI in creating concept imagery streamlines the design process and fosters a more informed and collaborative approach to architectural design.

Phase 3: Design Development (DD)

The design development phase is the third stage in the architectural design process. It entails the creation of more detailed and refined versions of the building design developed during the schematic design phase. In this phase, architects and designers work on elaborating the preliminary concepts and transforming them into comprehensive plans. During the design development phase, architects develop detailed floor plans, elevations, and sections, specifying materials and components to be used in the construction (Seyman Guray & Kismet, 2023). This includes making decisions on finishes, fixtures, and other essential design elements. The goal is to refine the initial design concept and ensure its feasibility and practicality for construction.

Moreover, architects start integrating the design development information into BIM software like Revit. This BIM integration allows for a more comprehensive and collaborative approach to design, enabling seamless coordination among the various design disciplines involved in the project.

AI Uses in Design Development

- **Schematic Plan Development:** AI uses generative design algorithms to create architectural floor plans based on input parameters like building size, room count, and desired layout. The AI system evaluates and refines the generated designs based on efficiency, functionality, and aesthetics. Additionally, AI algorithms can optimize existing floor plans by rearranging layouts to enhance efficiency and functionality using techniques like evolutionary algorithms or simulated annealing. The architectural design should be sufficiently thorough and detailed by the end of the design development phase for different stakeholders, such as clients, engineers, and contractors, to grasp the technical details and design intent of the project. The subsequent step, known as construction documentation, in which final construction documents are created based on design development work, is now set up for success.
- **3D Modeling/BIM:** AI can create detailed 3D models of buildings and environments, providing architects with powerful visualization and evaluation tools for design options. These AI-generated models offer a comprehensive representation of the architectural design, enabling architects to gain deeper insights into the spatial layout, aesthetics, and functionality of the proposed structures. By harnessing AI's capabilities, architects can efficiently explore and assess various design iterations, enhancing the decision-making process and ultimately leading to more informed and innovative architectural solutions (H. Wang et al., 2023).

Phase 4: Planning Application

The planning and drawing review phase hold significant importance in the architectural process, as it involves the critical step of evaluating and gaining approval for the design plans of a building or construction project. During this stage, architects present the suggested design to the appropriate authorities and parties, such as the neighborhood planning and building departments and other relevant organizations. This phase's main goal is to make that the suggested design conforms with all applicable laws, rules, and directives while also fulfilling the client's particular needs and specifications. Based on comments obtained throughout the review process, the architect might need to make changes to the design to ensure consistency with the defined standards and resolve any issues that the reviewing parties may have highlighted. Before the project is put into action, this phase is essential for ensuring that the design is both legally and practically workable.

AI Uses in Planning Application

Text-based AI, to its applications in the Pre-Design phase, can aid as a virtual assistant by managing scheduling, emails, documents, and monitoring project progress. This AI-driven support optimizes administrative tasks, allowing architects to focus on core design activities and enhancing overall project efficiency.

Phase 5: Construction Documentation (CD)

The construction document phase encompasses the creation of detailed drawings and specifications necessary for building the project. Known as "working drawings", these documents provide comprehensive information for contractors and tradespeople to understand and execute the architect's vision. Included are floor plans, sections, elevations, and details illustrating layout, size, and materials. Specifications for materials, as well as details of electrical and mechanical systems, may also be incorporated. Collaboration with engineers and contractors ensures accuracy and completeness (Shah Ali et al., 2008). Approved documents are used for bidding and serve as the foundation for project construction.

AI Uses in Construction Documents

- **Specification Writing:** AI can leverage data from past projects to analyze patterns and predict requirements for future endeavors. Architects and engineers can improve the precision and thoroughness of project specifications by using historical data (Wu et al., 2023). This data-driven methodology drastically cuts the time needed for project planning and design while also optimizing the decision-making process. AI's capability to analyze vast datasets empowers professionals to make informed and efficient decisions, streamlining the entire project development lifecycle.

Phase 6: Construction Administration (CA)

The construction administration phase serves as the concluding stage in the architectural process, wherein the architect takes on the responsibility of overseeing the construction of the building to ensure that it aligns with the approved design plans and specifications. During this phase, the architect is actively engaged in monitoring the progress and compliance of the construction activities. One of the primary activities in the construction administration phase is conducting regular site visits. The architect visits the construction site at predetermined intervals to observe the ongoing work, assess its quality, and verify that it adheres to the approved design and construction documents. By being physically present at the site, the architect can identify any deviations or discrepancies and promptly address them to maintain the project's alignment with the intended design. Furthermore, the architect plays a crucial role in reviewing and approving shop drawings and material samples submitted by contractors and suppliers (Chidambaram et al., 2012). Shop drawings are detailed representations of components, such as structural elements, mechanical systems, and finishes, prepared by the contractors. The architect examines these drawings to ensure that they accurately reflect the design intent and meet the necessary standards and codes. Similarly, the architect evaluates material samples to ensure they match the specified quality and aesthetic requirements. Throughout the construction administration phase, the architect acts as a mediator between the various stakeholders involved in the project. In the event of construction-related issues or conflicts, the architect facilitates communication and collaboration to find suitable resolutions. This can involve working closely with contractors, subcontractors, and the client to address challenges promptly and effectively. The overarching goal of the construction administration phase is to achieve the design intent as envisioned during the earlier stages of the architectural process. By closely monitoring the construction activities and maintaining strict adherence to the approved plans, the architect ensures that the final built environment aligns with the initial design concept and meets the client's needs and requirements (Ashworth et al., 2019).

AI Uses in Construction Administration

- **Punchlist Completion/Site Observations:** AI can use data from previous projects to anticipate if proposed initiatives will be feasible, including any potential delays or cost overruns. AI can spot patterns and trends that can point to difficulties or risks in the proposed project by using previous data and machine learning algorithms. Architects, engineers, and project managers can take preemptive measures to address possible problems and make better informed decisions thanks to this data-driven approach, which ultimately improves project planning and execution (Bibri, 2019).

AI-assisted Building Design: Optimization and Energy Efficiency

Data centers play a vital role in supporting our digital infrastructure, but their substantial energy consumption raises sustainability concerns. The data center sector is increasingly embracing AI as a potent tool to improve energy efficiency during construction and operation to address this issue (Ahmad et al., 2022). Since AI enables data center managers to intelligently optimize many parts of the facility, such as cooling systems, power utilization, and workload distribution, it plays a vital role in promoting energy-efficient practices.

During the construction phase, AI can assist in designing energy-efficient layouts and selecting eco-friendly materials. Moreover, AI-driven simulations can model and predict the data center's energy performance under different conditions, helping to identify potential areas for improvement.

Once operational, AI continuously monitors and optimizes energy consumption by dynamically adjusting cooling and power allocation based on real-time data and workload demands (Hameed et al., 2016). AI's predictive capabilities enable proactive maintenance, minimizing downtime and reducing energy wastage.

By harnessing AI's capabilities, data centers can significantly reduce their environmental impact, cutting energy usage and associated carbon emissions. This approach not only aligns with sustainability goals but also enhances overall performance and cost-effectiveness, making AI an indispensable tool for driving sustainable practices in the data center industry.

AI-driven Energy Efficiency: An Overview

In the realm of data center construction, AI plays a pivotal role by employing sophisticated algorithms and ML techniques to analyze extensive datasets, detect patterns, and make informed decisions. AI's integration facilitates the optimization of energy efficiency at all stages of the data center construction process. By joining AI's capabilities, data center operators can intelligently design layouts, select sustainable materials, and predict energy performance through simulations (Katal et al., 2023). Moreover, AI continuously monitors operational processes, dynamically adjusting cooling and power allocation based on real-time data and workload demands. This data-driven approach empowers data centers to achieve higher energy efficiency, reduce environmental impact, and maximize overall performance as shown in Table 7.2.

Intelligent Site Selection

AI algorithms are instrumental in analyzing geographical and environmental data to identify optimal locations for data centers. By considering factors such as climate conditions, proximity to renewable energy sources, and the existing local energy infrastructure, AI can strategically minimize energy consumption and maximize the utilization of sustainable resources (Chui et al., 2018). This data-driven approach empowers data center operators to make informed decisions, ensuring that the selected locations align with energy-efficient practices and contribute to overall environmental sustainability.

Table 7.2: AI applications for data center energy efficiency

Aspect	AI application
Intelligent site selection	Analyzing geographical and environmental data to identify optimal locations for data centers considering climate conditions, renewable energy sources, and existing local energy infrastructure.
Smart design and layout	Using AI-based design tools to optimize data center layouts by analyzing airflow, cooling requirements, and equipment placement for enhanced thermal management and reduced energy waste.
Predictive analytics	Harnessing historical and real-time data to make accurate predictions about energy consumption patterns, enabling proactive decision-making for optimal cooling settings and workload distribution.
Intelligent resource allocation	Monitoring and analyzing data center operations to dynamically adjust power distribution, workload balancing, and server utilization, ensuring efficient resource allocation.
Energy-efficient cooling systems	Continuously monitoring temperature fluctuations, airflow patterns, and server heat dissipation to make timely adjustments to cooling mechanisms for minimal energy consumption while maintaining optimal conditions.
Dynamic power management	Analyzing workload demands and dynamically adjusting power supply to optimize energy allocation during varying utilization levels, leading to cost savings and sustainable operation.
Continuous optimization	Equipped with machine learning capabilities, AI systems continually learn and adapt based on real-time data, identifying trends and anomalies for ongoing energy optimization and sustainability improvements.

Smart Design and Layout

AI-based design tools play a crucial role in assisting architects and engineers in creating energy-efficient data center layouts. These tools leverage AI algorithms to analyze factors such as airflow, cooling requirements, and equipment placement. By optimizing the design based on these considerations, AI minimizes energy waste and enhances thermal management, leading to more sustainable and cost-effective data center solutions, one of the reasons is the lack of theoretical understanding towards effective implementation (Hannan Qureshi et al., 2022). This integration of AI-driven design ensures that data centers operate at peak efficiency, contributing to reduced energy consumption and environmental impact.

Predictive Analytics for Energy Optimization

AI algorithms are capable of harnessing historical and real-time data to make accurate predictions about energy consumption patterns in data centers. By analyzing workload demand, temperature variations, and other relevant factors, AI empowers proactive decision-making to optimize energy usage. For instance, depending on real-time temperature data, AI may dynamically alter cooling settings, assuring efficient thermal management and minimizing energy loss. Additionally, AI has the ability to strategically divide workloads among servers to optimize resource usage and cut down on energy use (Abid et al., 2020). This data-driven strategy increases the effectiveness of data centers, resulting in significant energy savings and increased environmental sustainability.

Intelligent Resource Allocation

AI-driven systems play a crucial role in monitoring and analyzing data center operations, enabling optimal resource allocation to align with workload demands. Through dynamic adjustments in power distribution, workload balancing, and server utilization, AI minimizes energy waste and maximizes operational efficiency. By continuously analyzing real-time data, AI ensures that resources are utilized in the most efficient manner, leading to reduced energy consumption and enhanced performance for data centers (Kaur et al., 2020). This data-driven approach empowers data center operators to achieve significant energy savings while maintaining high levels of operational efficiency and sustainability.

Energy-efficient Cooling Systems

AI-powered algorithms play a crucial role in optimizing cooling systems within data centers. By continuously monitoring temperature fluctuations, airflow patterns, and server heat dissipation, AI can analyze real-time data to make informed decisions. This enables the system to make necessary adjustments to cooling mechanisms promptly, ensuring that energy consumption is minimized while maintaining optimal operating conditions. The data-driven approach of AI in managing cooling systems leads to significant energy savings and enhances the overall efficiency and sustainability of data center operations.

Dynamic Power Management

AI algorithms enable intelligent power management by analyzing workload demands and dynamically adjusting power supply accordingly. This data-driven approach ensures efficient energy allocation, reducing power waste during periods of low utilization and scaling up power during peak demands.

Continuous Optimization and Machine Learning

AI systems, equipped with ML capabilities, continually learn and adapt based on real-time data and feedback. This iterative learning process allows AI algorithms to identify trends, anomalies, and potential opportunities for further energy optimization within data centers. By analyzing vast datasets and recognizing patterns, AI-driven systems make ongoing improvements to enhance data center energy efficiency. This iterative learning approach ensures that the data center's energy management remains dynamic and continuously evolves to meet changing demands and improve sustainability over time.

AI is transforming the data center construction industry by promoting energy efficiency and sustainability. Leveraging intelligent site selection, optimized design, predictive analytics, resource allocation, and dynamic power management, AI plays a pivotal role in minimizing energy consumption and environmental impact. As AI technologies progress, their integration into data center construction becomes essential for achieving long-term sustainability goals.

AI-driven energy optimization not only benefits the environment but also provides significant cost savings and operational efficiencies for data center operators. Embracing AI empowers the data center industry to create a greener and more sustainable future, balancing technological advancement with environmental responsibility (Mariani et al., 2023).

Summary

The integration of AI in building design has transformative potential in the field of architecture, revolutionizing how architects conceive and create structures. AI technologies offer architects exciting opportunities to explore innovative and eco-friendly design solutions. By harnessing computational power and data-driven algorithms, AI can evaluate vast amounts of data, including building performance, environmental conditions, and user preferences, to inform the design process and improve building performance. This collaboration between human designers and AI fosters a symbiotic relationship, enhancing architects' creative capabilities and efficiency in the design process.

AI-assisted building design techniques include generative design, performance optimization, parametric design, and NLP. These techniques empower architects to generate and explore design alternatives, optimize building systems, automate repetitive tasks, and extract valuable insights from textual data.

AI can be integrated into the architectural design process in the following ways: Pre-Design (PD), Schematic Design (SD), Design Development (DD), Planning Application, Construction Documents (CD), Construction Administration (CA).

In data centers, AI promotes energy efficiency through intelligent site selection, optimized design, predictive analytics, dynamic power management, and continuous optimization using machine learning. It helps reduce environmental impact and operational costs while enhancing performance.

The integration of AI in building design raises ethical and societal concerns, prompting the need to strike a balance between AI capabilities and human creativity to ensure human-centered design outcomes.

In conclusion, AI-assisted building design offers immense potential for revolutionizing the architecture field, enriching creative capacities, enhancing design efficiency, and creating sustainable and user-centric structures. However, careful consideration of ethical implications is essential to ensure a human-centered and ethically responsible design process. Case studies of AI-assisted building design demonstrate its efficacy in design optimization, energy efficiency, smart cities, historic building preservation, construction management, and client communication. AI continues to reshape the architectural landscape, promising more profound impacts on how buildings are designed and cities are shaped.

In the conceptual and schematic stages of building design, a generic data flow model, Design Structure Matrix (DSM), and simulation models contribute to effective design management. Understanding and optimizing these early design phases are crucial for successful architectural projects and cost minimization.

References

Abid, A., Manzoor, M.F., Farooq, M.S., Farooq, U. and Hussain, M. (2020). Challenges and issues of resource allocation techniques in cloud computing. *KSII Transactions on Internet and Information Systems*, 14(7), 2815–2839. https://doi.org/10.3837/tiis.2020.07.005.

Ahmad, T., Zhu, H., Zhang, D., Tariq, R., Bassam, A., Ullah, F., AlGhamdi, A.S. and Alshamrani, S.S. (2022). Energetics systems and artificial intelligence: Applications of industry 4.0. *Energy Reports*, 8, 334–361. https://doi.org/https://doi.org/10.1016/j.egyr.2021.11.256.

Allam, Z. and Dhunny, Z.A. (2019). On big data, artificial intelligence and smart cities. *Cities*, 89, 80–91. https://doi.org/https://doi.org/10.1016/j.cities.2019.01.032.

Ammad, S., Alaloul, W.S., Saad, S., Altaf, M., Alawag, A.M. and Ali, M. (2021). Building Information Modeling (BIM) and Occupational Safety in Infrastructure Projects. *2021 International Conference on Data Analytics for Business and Industry (ICDABI)*, pp. 240–244. https://doi.org/10.1109/ICDABI53623.2021.9655832.

Ammad, S., Saad, S., Bashir, M.T., Qureshi, A.H., Altaf, M. and Rasheed, K. (2021). Construction Accidents via Integrating Building Information Modeling (BIM) with Emerging Digital Technologies: A Review. *2021 3rd International Sustainability and Resilience Conference: Climate Change*. https://doi.org/10.1109/IEEECONF53624.2021.9668066.

Ashworth, S., Tucker, M. and Druhmann, C.K. (2019). Critical success factors for facility management employer's information requirements (EIR) for BIM. *Facilities*, 37(1/2), 103–118. https://doi.org/10.1108/F-02-2018-0027.

Baduge, S.K., Thilakarathna, S., Perera, J.S., Arashpour, M., Sharafi, P., Teodosio, B., Shringi, A. and Mendis, P. (2022). Artificial intelligence and smart vision for building and construction 4.0: Machine and deep learning methods and applications. *Automation in Construction*, 141, 104440. https://doi.org/https://doi.org/10.1016/j.autcon.2022.104440.

Baldwin, A.N., Austin, S.A., Hassan, T.M. and Thorpe, A. (2010). Modelling information flow during the conceptual and schematic stages of building design. 17(2), 155–167. https://doi.org/10.1080/014461999371655.

Bibri, S.E. (2019). The anatomy of the data-driven smart sustainable city: Instrumentation, datafication, computerization and related applications. *Journal of Big Data*, 6(1), 59. https://doi.org/10.1186/s40537-019-0221-4.

Bölek, B., Tutal, O. and Özbaşaran, H. (2023). A systematic review on artificial intelligence applications in architecture. *Journal of Design for Resilience in Architecture and Planning*, 4(1), 91–104. https://doi.org/10.47818/DRARCH.2023.V4I1085.

Chidambaram, R., Narayanan, S.P. and Idrus, A.B. (2012). Construction delays causing risks on time and cost: A critical review. *The Australasian Journal of Construction Economics and Building*, 12(1), 37-57. https://search.informit.org/doi/10.3316/informit.119538752826363.

Chui, K.T., Lytras, M.D. and Visvizi, A. (2018). Energy sustainability in smart cities: Artificial intelligence, smart monitoring, and optimization of energy consumption. *Energies*, 11(11). https://doi.org/10.3390/en11112869.

Ding, Y., Ma, J. and Luo, X. (2022). Applications of natural language processing in construction. *Automation in Construction*, 136, 104169. https://doi.org/https://doi.org/10.1016/j.autcon.2022.104169.

Dwivedi, Y.K., Hughes, L., Ismagilova, E., Aarts, G., Coombs, C., Crick, T., Duan, Y., Dwivedi, R., Edwards, J., Eirug, A., Galanos, V., Ilavarasan, P.V., Janssen, M., Jones, P., Kar, A.K., Kizgin, H., Kronemann, B., Lal, B., Lucini, B., … Williams, M.D. (2021). Artificial Intelligence (AI): Multidisciplinary perspectives on emerging challenges, opportunities, and agenda for research, practice and policy. *International Journal of Information Management*, 57, 101994. https://doi.org/https://doi.org/10.1016/j.ijinfomgt.2019.08.002.

Hafez, F.S., Sa'di, B., Safa-Gamal, M., Taufiq-Yap, Y.H., Alrifaey, M., Seyedmahmoudian, M., Stojcevski, A., Horan, B. and Mekhilef, S. (2023). Energy efficiency in sustainable buildings: A systematic review with taxonomy, challenges, motivations, methodological aspects, recommendations, and pathways for future research. *Energy Strategy Reviews*, 45, 101013. https://doi.org/https://doi.org/10.1016/j.esr.2022.101013.

Hameed, A., Khoshkbarforoushha, A., Ranjan, R., Jayaraman, P.P., Kolodziej, J., Balaji, P., Zeadally, S., Malluhi, Q.M., Tziritas, N., Vishnu, A., Khan, S.U. and Zomaya, A. (2016). A survey and taxonomy on energy efficient resource allocation techniques for cloud computing systems. *Computing*, 98(7), 751–774. https://doi.org/10.1007/s00607-014-0407-8.

Hannan Qureshi, A., Alaloul, W.S., Wing, W.K., Saad, S., Ammad, S. and Musarat, M.A. (2022). Factors impacting the implementation process of automated construction progress monitoring. *Ain Shams Engineering Journal*, 13(6), 101808. https://doi.org/https://doi.org/10.1016/j.asej.2022.101808.

Hannan Qureshi, A., Salah Alaloul, W., Kai Wing, W., Saad, S., Ali Musarat, M., Ammad, S. and Farouk Kineber, A. (2023). Automated progress monitoring technological model for construction projects. *Ain Shams Engineering Journal*, 14(10), 102165. https://doi.org/https://doi.org/10.1016/j.asej.2023.102165.

Katal, A., Dahiya, S. and Choudhury, T. (2023). Energy efficiency in cloud computing data centers: A survey on software technologies. *Cluster Computing*, 26(3), 1845–1875. https://doi.org/10.1007/s10586-022-03713-0.

Kaur, K., Garg, S., Kaddoum, G., Bou-Harb, E. and Choo, K.-K.R. (2020). A big data-enabled consolidated framework for energy efficient software defined data centers in IoT setups. *IEEE Transactions on Industrial Informatics*, 16(4), 2687–2697. https://doi.org/10.1109/TII.2019.2939573.

Lluís Zamora Mestre, D., Coloma Picó, D. and Reza Zaker Hossein, M. (2019). BIM implementation in architectural practices: Towards advanced collaborative approaches based on digital technologies. *TDX* (Tesis Doctorals En Xarxa).

Mariani, M.M., Machado, I. and Nambisan, S. (2023). Types of innovation and artificial intelligence: A systematic quantitative literature review and research agenda. *Journal of Business Research*, 155, 113364. https://doi.org/https://doi.org/10.1016/j.jbusres.2022.113364.

McKenna-Cress, P. and Kamien, J. (2013). *Creating Exhibitions: Collaboration in the Planning, Development, and Design of Innovative Experiences*. John Wiley & Sons.

Mehmood, M.U., Chun, D., Zeeshan, Han, H., Jeon, G. and Chen, K. (2019). A review of the applications of artificial intelligence and big data to buildings for energy-efficiency and a comfortable indoor living environment. *Energy and Buildings*, 202, 109383. https://doi.org/https://doi.org/10.1016/j.enbuild.2019.109383.

Pan, Y. and Zhang, L. (2023). Integrating BIM and AI for smart construction management: Current status and future directions. *Archives of Computational Methods in Engineering*, 30(2), 1081–1110. https://doi.org/10.1007/s11831-022-09830-8.

Parihar, V., Malik, A., Bhawna, Bhushan, B. and Chaganti, R. (2023). From smart devices to smarter systems: The evolution of Artificial Intelligence of Things (AIoT) with characteristics, architecture, use cases and challenges. *In:* B. Bhushan, A.K. Sangaiah and T.N. Nguyen (Eds.), *AI Models for Blockchain-Based Intelligent Networks in IoT Systems: Concepts, Methodologies, Tools, and Applications*, pp. 1–28. Springer International Publishing. https://doi.org/10.1007/978-3-031-31952-5_1

Qureshi, A.H., Alaloul, W.S., Manzoor, B., Musarat, M.A., Saad, S. and Ammad, S. (2020). Implications of machine learning integrated technologies for construction progress detection under industry 4.0 (IR 4.0). *2020 Second International Sustainability and Resilience Conference: Technology and Innovation in Building Designs (51154)*, 1–6. https://doi.org/10.1109/IEEECONF51154.2020.9319974.

Rasheed, K., Ammad, S., Said, A.Y., Balbehaith, M., Oad, V.K. and Khan, A. (2022). Application of Smart IoT Technology in Project Management Scenarios. *2022 International Conference on Data Analytics for Business and Industry, ICDABI 2022*. https://doi.org/10.1109/ICDABI56818.2022.10041674.

Rigillo, M. (2023). Neil Leach architecture in the age of Artificial Intelligence: An introduction to AI for architects. *Techne*, 25, 272–273. https://doi.org/10.36253/techne-14769.

Saad, S., Alaloul, W.S., Ammad, S. and Qureshi, A.H. (2022). A qualitative conceptual framework to tackle skill shortages in offsite construction industry: A scientometric approach. *Engineering, Construction and Architectural Management*, 29(10), 3917–3947. https://doi.org/10.1108/ECAM-04-2021-0287.

Saad, S., Alaloul, W.S., Rasheed, K. and Ammad, S. (2022). Modeling and simulation of construction cyber-physical systems. *In: Cyber-Physical Systems in the Construction Sector*, pp. 88–110. CRC Press.

Sacks, R., Girolami, M. and Brilakis, I. (2020). Building information modeling, artificial intelligence and construction tech. *Developments in the Built Environment*, 4, 100011. https://doi.org/10.1016/J.DIBE.2020.100011.

Seyman Guray, T. and Kismet, B. (2023). Applicability of a digitalization model based on augmented reality for building construction education in architecture. *Construction Innovation*, 23(1), 193–212. https://doi.org/10.1108/CI-07-2021-0136.

Shah Ali, A., Rahmat, I. and Hassan, H. (2008). Involvement of key design participants in refurbishment design process. *Facilities*, 26(9/10), 389–400. https://doi.org/10.1108/02632770810885742.

Syed, S., Alaloul, W.S. and Ammad, S. (2022). Fundamentals of construction project management. *In:* D.W.S. Alaloul (Ed.), *Cyber-Physical Systems in the Construction Sector* (1st Edn.), p. 21, CRC Press. https://doi.org/10.1201/9781003190134-1.

Wang, H., Fu, T., Du, Y., Gao, W., Huang, K., Liu, Z., Chandak, P., Liu, S., Van Katwyk, P., Deac, A., Anandkumar, A., Bergen, K., Gomes, C.P., Ho, S., Kohli, P., Lasenby, J., Leskovec, J., Liu, T.-Y., Manrai, A., … Zitnik, M. (2023). Scientific discovery in the age of artificial intelligence. *Nature*, 620(7972), 47–60. https://doi.org/10.1038/s41586-023-06221-2.

Wang, K., Ying, Z., Goswami, S.S., Yin, Y. and Zhao, Y. (2023). Investigating the role of Artificial Intelligence technologies in the construction industry using a Delphi-ANP-TOPSIS Hybrid MCDM concept under a fuzzy environment. *Sustainability*, 15(15). https://doi.org/10.3390/su151511848.

Wu, J., Guo, Y., Gao, C. and Sun, J. (2023). An automatic text generation algorithm of technical disclosure for catenary construction based on knowledge element model. *Advanced Engineering Informatics*, 56, 101913. https://doi.org/https://doi.org/10.1016/j.aei.2023.101913.

Yadav, S., Prakash, A., Arora, M. and Mittal, A. (2023). Digital transformation: Exploring cornerstones for construction industry. *Kybernetes* (ahead-of-print). https://doi.org/10.1108/K-05-2023-0895.

CHAPTER

8

AI in Fabrication and Construction

Muhammad Haris, Syed Saad, Syed Ammad and Kumeel Rasheed

Introduction

Artificial intelligence (AI) has sparked a revolution across various industries, captivating the global audience. The profound impact of AI can be observed in the way we interact with technology, where digital assistants such as Siri and Alexa, alongside advanced AI tools like ChatGPT, have brought about dramatic changes. While AI has found applications in diverse fields, one sector that stands to immensely benefit from its adoption is the construction industry, which has historically been one of the least digitalized sectors worldwide. Construction sites could become more efficient, better in terms of quality and performance, and safer thanks to the incorporation of AI.

Every phase of construction is being rapidly transformed by AI and machine learning (ML) technology, from design and planning to operations and asset management. Researchers' current focus areas include automated building progress detection and IR 4.0. However, as an evolving idea, the application of the IR 4.0 theme for progress detection technology requires unique considerations (Abdul Hannan Qureshi et al., 2020). These technological advancements are generating significant improvements in terms of time and cost-effectiveness, particularly in the planning and designing sub-segment.

The implementation of AI in construction management enhances safety, speed, accuracy, and efficiency of operations. With AI-powered construction solutions, companies can proactively monitor job sites to identify potential hazards and address critical issues to prevent accidents. ML in construction enables the analysis of vast amounts of data, optimizing designs for energy efficiency and cost savings. Robotics process automation (RPA) plays a crucial role in improving construction speed and quality.

In essence, AI is revolutionizing the construction industry, revolutionizing how businesses conceive, execute, and manage projects. As the adoption of AI continues to expand, companies find it increasingly essential to adapt to this emerging technology. Leading the charge in AI adoption within the construction sector is the European market, propelling growth in the global construction industry. By embracing digitization, all stakeholders, including contractors, owners, and service providers, can reap the benefits of this technological shift. As the industry is still in its early stages of adopting AI, companies that embrace and upgrade their technology stand to gain a competitive advantage.

AI has become an indispensable component of the modern engineering and construction industry. Its application in construction helps tackle the sector's greatest challenges, including cost, safety, and project overruns. Throughout the construction project lifecycle, from inception and design to bidding, financing, transportation, operation, and asset management, AI plays a vital role.

The handling and execution of work has recently undergone a significant change as a result of the use of new technology. Infrastructure projects are increasingly using building information modeling (BIM) to increase the efficiency of a number of jobs (Ammad, Alaloul, et al., 2021). As AI continues to be leveraged for construction, the industry envisions a future where collaborative robots (cobots) and robots work alongside human workers. Robots take over tasks that can be automated, while cobots operate autonomously or with limited guidance. This arrangement accelerates the construction process, reduces costs, minimizes injuries, enhances efficiency, and facilitates informed decision-making.

Moreover, AI in construction is poised to bring about changes in business models, significantly reducing costly errors and making building operations more efficient. In the construction industry, the policymakers tend toward economic sustainability. Whereas in the construction industry the associated future cost inflicts barriers on project sustainability (Altaf et al., 2020). Therefore, construction companies' leaders are encouraged to invest in AI where it can have the maximum impact, based on their specific needs. Early adopters of this digital transformation are certain to drive growth and gain a competitive edge, setting the direction for the industry and reaping the maximum benefits.

In conclusion, AI in construction presents a vast array of opportunities to enhance various aspects of the industry. By capitalizing on the capabilities of machine learning, robotics, drones, and other AI-powered solutions, construction companies can revolutionize their operations. This transformative approach empowers them to optimize processes, enhance safety measures, boost productivity, automate tasks, and base decisions on data-driven insights, leading to a more efficient and sustainable construction ecosystem. The key to shaping the future of the construction industry lies in its adeptness at embracing and effectively leveraging AI technologies.

AI-based Fabrication Techniques

The manufacturing industry depends heavily on fabrication procedures, which include a variety of techniques like additive manufacturing (3D printing), subtractive manufacturing (CNC machining), and shaping operations as shown in Figure 8.1. Precision, speed, and material utilization are all constrained by traditional fabrication methods. These procedures have been transformed by AI, which has opened up transformative possibilities that have fundamentally altered how components are designed, produced, and assembled. This study aims to give information on the state of fabrication processes based on AI, their applications, and their potential to transform the industrial sector (Zanoni et al., 2019).

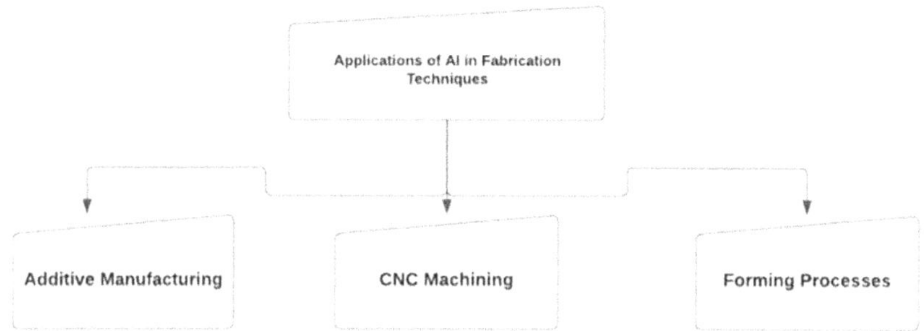

Figure 8.1: Applications of AI in fabrication techniques.

The Significance of AI for the Future of Manufacturing

In the quest for sustained growth, cost reduction, risk mitigation, and enhanced production efficiency, manufacturers are seeking innovative solutions. To achieve these goals, the adoption of Fourth Industrial Revolution (4IR) technologies, particularly AI-based and ML-powered innovations, is becoming increasingly critical.

AI tools have the capacity to process and analyze vast volumes of data from the production floor, enabling the identification of patterns, analysis of consumer behavior, real-time detection of production process anomalies, and more. These tools offer manufacturers comprehensive visibility into all aspects of manufacturing operations across diverse geographical locations. Leveraging ML algorithms, AI-powered systems can continuously learn, adapt, and improve their performance, rendering them indispensable in the rapidly digitized post-pandemic landscape.

The implementation of AI in manufacturing also enables predictive analytics, which proves invaluable in addressing operational challenges and disruptions to supply chains and the workforce.

Furthermore, AI offers a plethora of additional benefits to the manufacturing sector, including predictive maintenance, which effectively reduces unplanned downtime. The utilization of advanced manufacturing technologies, such as 3D printers and robots, in near-shore facilities helps lower labor costs and ensures resilience despite supply chain disruptions. Additionally, AI-enabled generative design enables the creation of optimal and efficient design solutions, thereby minimizing waste in the manufacturing process.

In conclusion, the integration of AI in manufacturing processes holds significant promise for enhancing productivity, reducing costs, and mitigating risks as shown in Table 8.1. As the manufacturing world evolves, AI's role in driving innovation and efficiency becomes increasingly vital to achieving long-term success and sustainable growth.

Table 8.1: Applications of AI in fabrication techniques

Fabrication technique	Application of AI in fabrication	Advantages and contributions
Additive manufacturing	AI techniques aid in streamlining production processes, optimizing design, and achieving higher resource utilization efficiency.	Improved production efficiency and resource utilization. Enhanced design optimization and reduced waste. Predictive maintenance for reduced unplanned downtime. Automation of complex tasks for increased efficiency.
CNC machining	AI enables autonomous programming of CNC machines and optimizes tool path selection.	Significant reduction in programming time and human errors. Improved machining accuracy and surface quality. Cost and time optimization in CNC machining processes.
Forming processes	AI facilitates the analysis of Incremental Sheet Forming (ISF) and mapping of nonlinear behavior.	Faster and more efficient analysis compared to traditional methods. Reduction in the number of experiments needed. Enhanced understanding of relationships between ISF parameters and final product attributes.

AI in Additive Manufacturing

The incorporation of AI into additive manufacturing (AM) is gaining widespread recognition for its capacity to revolutionize diverse sectors. Despite its promising potential, there is a pressing requirement to improve the efficiency and dependability of additive manufacturing in product development. AM, commonly referred to as 3D printing, has gained extensive adoption across manufacturing processes (Wang et al., 2020). This research aims to explore the integration of AI techniques in AM and address existing challenges to fully unlock its transformative impact on the manufacturing industry. By harnessing

AI's capabilities, AM holds the promise of streamlining production processes, optimizing design, and achieving higher resource utilization efficiency, thereby paving the way for innovative advancements in the manufacturing field (Wang et al., 2020). This paper investigates the current progress of AI in AM, identifies research gaps, and proposes potential solutions to drive the future of additive manufacturing towards enhanced efficiency and productivity.

AI techniques play a pivotal role in advancing the development of AM products, particularly through the utilization of intelligent agents. These agents are instrumental in addressing challenges at various stages of the AM process, encompassing product design, process design, and production phases. Employing sophisticated AI algorithms, these agents proficiently analyze vast data sets, make informed decisions, and optimize the AM process to achieve desired outcomes. By integrating AI into AM, automation of complex tasks becomes feasible, thereby elevating overall manufacturing process efficiency (Wang et al., 2020).

Current Development and Research Gaps

Due to their effectiveness and efficiency, specialists in the construction sector have praised the development of digital data-acquisition technologies in monitoring operations. But there has been reluctance to accept these technologies due to a lack of theoretical understanding of their successful implementation (Hannan Qureshi et al., 2023). The present state of AI-enabled AM product development is actively under exploration, with researchers investigating the efficacy of AI techniques in diverse areas, including product design, process optimization, and quality control within the AM domain. Despite these advancements, certain research gaps persist and necessitate attention. Notably, there is a need for the development of more efficient and comprehensive intelligent agents capable of addressing multifaceted challenges in AM (Wang et al., 2020). Moreover, seamless integration of AI with emerging technologies, such as cloud-edge computing, poses an area of exploration to unlock further potential in AI-enhanced AM product development. Addressing these research gaps will be crucial in harnessing the full potential of AI-driven additive manufacturing and propelling the field towards greater efficiency and innovation (Wang et al., 2020).

AI in CNC Machining

AI has brought about a transformative impact on the domain of Computer Numerical Control (CNC) machining, significantly enhancing its efficiency and precision. CNC machines are extensively utilized in manufacturing industries for machining parts, and the selection of an optimal tool path holds paramount importance in optimizing the machining process. Historically, tool path selection relied on manual programming and data from machining handbooks, resulting in time-consuming processes prone to errors (Author et al., 2014).

Recent advancements in AI and advanced computer-aided manufacturing (CAM) software have enabled the autonomous programming of CNC machines.

This revolutionary development significantly reduces programming time and minimizes the potential for human errors. Optimizing tool path selection is crucial in the world of CNC machining, and AI techniques have been used to make this happen. It has been effectively implemented to improve tool path selection in CNC machining processes to use strategies including Genetic Algorithms (GA), Artificial Neural Networks (ANNs), Artificial Immune Systems (AIS), Ant Colony Optimization (ACO), and Particle Swarm Optimization (PSO).

The primary objective of AI-based tool path optimization is to attain high machining accuracy, minimize production time and cost, improve surface quality, and enhance overall productivity. Researchers have focused on various parameters, such as the reduction of machining time, cost optimization, optimization of tool travel path, and enhancement of surface quality.

By leveraging AI methods, CNC machining attains elevated levels of efficiency and precision, thus fostering improved manufacturing processes across diverse industries. The integration of AI technologies in CNC machining presents significant opportunities to streamline operations and enhance the overall productivity and quality of the manufacturing process.

AI in Forming Processes

AI has found widespread applications across various industries, encompassing manufacturing, healthcare, sports, and finance, enabling the modeling of nonlinearities and facilitating reliable predictions. Within the manufacturing domain, AI has been leveraged to enhance processes, drive cost reduction, and elevate reliability. An area where AI has shown particular promise is Incremental Sheet Forming (ISF), an innovative manufacturing process involving the step-by-step incremental feed of a sheet metal or polymer blank using a CNC machine (Harfoush et al., 2021).

The quality of the final product in ISF is influenced by a multitude of parameters pertaining to the forming process, material properties, and geometric factors. Traditionally, the study of relationships between these parameters and the attributes of the final product relied on analytical, experimental, and numerical techniques.

However, these conventional methods often encounter inefficiencies due to the inherent nonlinearities and complexities involved. Additionally, the process is time-consuming, and simulations demand extensive computational resources.

To surmount these challenges, researchers have turned to AI techniques for the analysis of ISF. AI models offer effective mapping of the process's nonlinear behavior, reducing the need for a large number of experiments, and yielding faster and more efficient outcomes compared to traditional approaches.

The application of AI in ISF research aims to enhance the comprehension of relationships between ISF parameters and final product attributes, while also identifying potential avenues for future investigation (Harfoush et al., 2021).

AI-assisted Construction Planning and Management

By offering insightful data and increasing effectiveness, AI has the potential to change construction planning and administration.

Construction managers may foresee and adapt to upcoming possibilities and difficulties with the use of AI-generated scenarios, which will help them make better decisions. But there are additional organizational and ethical issues that need to be taken into account when using AI in building planning (Spaniol & Rowland, 2023).

AI-generated Scenarios for Strategic Planning: Their Benefits and Challenges

Benefits of AI-generated Scenarios for Strategic Planning

The utilization of AI-generated scenarios holds significant promise in providing invaluable insights to managers, enabling them to effectively anticipate and adapt to forthcoming challenges and opportunities. By exploring a diverse range of potential future outcomes, these AI-generated scenarios can enhance the strategic planning process and expand the array of strategic options available to organizations. Integrating such scenarios into the planning process is particularly beneficial for organizations operating in turbulent environments and those striving to bolster their strategic capabilities. As a result, AI-generated scenarios can play a crucial role in empowering managers to make well-informed decisions and navigate uncertainties with greater confidence and foresight (Abbasi et al., 2022). The integration of AI-driven scenario analysis in strategic planning equips organizations with a proactive and versatile approach to address ever-changing market dynamics, ultimately fostering adaptability and resilience in the face of dynamic business landscapes (Spaniol & Rowland, 2023).

Challenges of AI-generated Scenarios for Strategic Planning

The integration of artificial intelligence-generated scenarios in decision-making processes brings to the forefront ethical concerns, particularly regarding biases and discrimination that may arise from the underlying data and algorithms (Bahroun et al., 2023). Users must be conscientious about these ethical implications and take proactive measures to mitigate potential biases. Regular evaluation and updating of the data and algorithms are essential to ensure that the scenarios generated remain fair, impartial, and inclusive.

Incorporating AI-generated scenarios may necessitate organizational adjustments to accommodate this emerging technology effectively. Creating new procedures and developing the necessary skills within the workforce are vital steps to ensure seamless integration and utilization of AI-generated scenarios. This may involve training personnel on AI technologies, data analysis, and ethical considerations. Moreover, fostering a culture of transparency and accountability within the organization is essential to promote responsible use and interpretation of AI-generated scenarios. Implementing these changes requires dedication of both time and financial resources. Organizations must allocate sufficient time for training, testing, and refining the AI systems to ensure accuracy and reliability of the generated scenarios. Adequate financial investments are also required to procure cutting-edge AI technologies, provide training programs, and establish a robust infrastructure for data collection, processing, and analysis.

Furthermore, fostering collaboration and buy-in from all relevant stakeholders is critical to the successful integration of AI-generated scenarios. Engaging various departments, management levels, and external partners in the decision-making process ensures a holistic understanding of the technology's capabilities and limitations (Ammad, Saad et al., 2021). Transparent communication and open dialogue regarding the implications and benefits of AI-generated scenarios can facilitate widespread acceptance and support.

By adopting a thoughtful and proactive approach, organizations can harness the power of AI-generated scenarios while responsibly addressing ethical concerns and optimizing their decision-making capabilities (Patil & Shankar, 2023). Embracing this transformative technology will require a combination of technical expertise, organizational adaptability, and a commitment to ethical best practices, paving the way for more informed and responsible decision-making in a rapidly evolving AI-driven landscape.

Best Practices for Using AI-generated Scenarios

AI-generated scenarios should be used as part of a bigger strategic planning process rather than being employed independently. It is essential for users to carefully assess how AI-generated scenarios align with their overarching strategies and their potential to enhance decision-making and action planning. Regularly engaging in discussions and critical evaluations of these future scenarios facilitated by AI technologies is recommended to foster "futures literacy" or "futures consciousness" among personnel, particularly managerial staff. This proactive approach enables individuals to develop a deeper understanding of potential future developments and empowers them to make more informed and strategic decisions in response to evolving circumstances. Responsible and effective use of AI-generated scenarios is key to their value in strategic planning (Gil et al., 2020).

In conclusion incorporating AI-generated scenarios into strategic planning practices can bring benefits in terms of insights and expanded strategic options. However, it is important to address ethical concerns, invest in organizational changes, and integrate these scenarios into the broader planning process. AI-generated scenarios can be a valuable tool for strategic planning if used responsibly and effectively.

Best Practices for Incorporating AI-generated Scenarios into the Strategic Planning Process

Increase "Futures Literacy" of Employees

To effectively integrate AI-generated scenarios into strategic planning practices, it is highly advisable to focus on enhancing the "futures literacy" or "futures consciousness" of employees, particularly managerial staff. This can be achieved by encouraging regular and meaningful discussions and analyses of the future scenarios generated by AI tools.

Fostering a culture of futures literacy entails developing a deeper understanding of the potential implications and impacts of AI-generated scenarios on the organization's strategic objectives. Employees can learn a lot about numerous prospective futures, potential obstacles, and new possibilities by having continual talks and probing these scenarios.

The managerial staff's participation is especially important because they are vital in developing and putting into practice organizational strategy. Encouraging their active participation in analyzing AI-generated scenarios can lead to more informed and nuanced strategic planning, considering a broader range of possible outcomes and responses to future challenges.

Strategic planning practices should go beyond occasional strategy development sessions and instead incorporate the regular examination of AI-generated scenarios (Williams & Yampolskiy, 2021). This ensures that strategic decisions are continuously informed by up-to-date and relevant insights, leading to greater adaptability and agility in response to dynamic market conditions and emerging trends.

Consider Ethical and Organizational Considerations

The integration of AI-generated scenarios into decision-making processes necessitates careful attention to both ethical and organizational aspects. Managers must be diligent in acknowledging and addressing potential biases and discrimination that could arise from these scenarios and take proactive steps to mitigate them. To achieve this, regular review and updates of the data and algorithms employed in generating the scenarios become crucial.

Ethical considerations are of paramount importance when incorporating AI-generated scenarios into strategic planning and decision-making. AI algorithms are trained on vast datasets, and if these datasets contain biased or discriminatory information (Norori et al., 2021), it can inadvertently lead to biased outcomes in the generated scenarios. Such biases can reinforce existing inequalities or unfairly disadvantage certain groups.

To mitigate these risks, managers should be vigilant in scrutinizing the datasets used to train AI algorithms. They must ensure that the data is representative and unbiased, or take corrective actions to remove any biased patterns. Moreover, ongoing monitoring and evaluation of the generated scenarios can help identify and rectify any emerging biases, ensuring fairness and inclusivity.

Organizational considerations also play a crucial role in the successful integration of AI-generated scenarios. Regularly reviewing the data and algorithms used in the AI models is essential to maintain their accuracy and relevance. As business environments evolve, the underlying assumptions and factors driving the scenarios may change, making it necessary to update the AI models to reflect these shifts.

Furthermore, organizations should foster a culture of transparency and openness regarding the use of AI-generated scenarios. Managers must be vigilant in addressing potential biases and discrimination, regularly reviewing and updating the data and algorithms, and promoting transparency in their usage. By adhering to these principles, organizations can harness the full potential of AI-generated scenarios while ensuring fairness and responsible decision-making.

Prepare for Organizational Changes

The seamless integration of AI-generated scenarios into the strategic planning process may necessitate organizational adjustments, including the creation of new procedures and the development of relevant skills. Such adjustments are vital to effectively leverage the insights provided by AI-generated scenarios and to ensure their successful application in decision-making.

To effectively incorporate AI-generated scenarios, organizations need to establish new procedures that facilitate the collection, analysis, and utilization of data to generate meaningful scenarios (Paschen et al., 2020). This may involve the integration of AI technologies into existing data management systems or the implementation of new data gathering methods to ensure the availability of relevant and accurate information for scenario generation.

Additionally, developing the necessary skills within the organization is crucial. This involves providing training and educational opportunities for employees to enhance their understanding of AI technologies and data analysis. Employees that are given the knowledge and abilities to understand and use AI-generated situations are better able to act on the information presented and make wise judgments.

The acceptance and collaboration of all organization stakeholders are necessary for the successful implementation of these changes. Clear communication regarding the advantages of AI-generated scenarios and the value they provide to the strategic planning process is necessary to ensure that all pertinent parties are on board. Transparency and open communication with stakeholders can support a collaborative implementation strategy by addressing any concerns or resistance (Riege & Lindsay, 2006).

It is crucial to realize that incorporating AI-generated scenarios into the process of strategic planning is a time-consuming procedure that may call for a sizable time and money commitment. Organizations should be prepared to devote the necessary amount of time, money, and human capital in order to successfully assist the transformation.

In conclusion, businesses may need to undertake organizational adjustments, develop specialized skills, and assure the support of all stakeholders to fully realize the potential of AI-generated scenarios in strategic planning. Embracing these changes and dedicating the necessary resources will enable organizations to harness the benefits of AI-generated scenarios and make well-informed decisions for their future success.

Integrate AI-generated Scenarios into Overall Strategy

The integration of AI-generated scenarios into the broader strategic planning process is of paramount importance, rather than utilizing them in isolation. It is crucial to carefully consider how these scenarios fit with the overarching strategic goals and how they might help with decision-making and action planning.

Decision-makers can learn a lot about a range of potential future difficulties and possibilities by adding AI-generated scenarios. This enables them to create flexible and resilient plans that can adjust to shifting conditions. AI-generated scenarios offer a multitude of possible outcomes, each with its associated probabilities. This rich diversity of scenarios aids in identifying potential risks and uncertainties that an organization may encounter. Moreover, it facilitates the development of contingency plans to address unforeseen challenges.

An integral aspect of strategic success lies in precisely considering these AI-generated scenarios. Organizations can use the potential of these circumstances to take proactive actions consistent with their long-term objectives. Using AI-generated scenarios as part of the strategic planning process enables organizations to be proactive and well-prepared for a number of probable futures. In conclusion, the integration of AI-generated scenarios into the strategic planning process offers immense benefits to organizations. By aligning these scenarios with the overall strategy and leveraging their insights, decision-makers can develop adaptive and robust strategies. Furthermore, the diverse range of scenarios and their associated probabilities assist in risk identification and contingency planning. It is through the thoughtful consideration and incorporation of these AI-generated scenarios that organizations can pave the way for strategic success in an ever-changing and dynamic business environment.

AI in the Construction Industry: Application in Scenario Planning

AI has the potential to revolutionize the construction industry, including its application in scenario planning. By leveraging AI technologies, construction companies can generate scenarios that explore possible future outcomes and assist in strategic planning.

Benefits of AI-generated Scenarios

The utilization of AI-generated scenarios in strategic planning holds significant promise for construction companies. These scenarios provide insightful information that can help managers anticipate and respond to upcoming opportunities and problems, enabling them to make wise decisions.

The capacity of AI-generated scenarios to generate a wide range of options on numerous themes pertinent to the construction business is one of their standout benefits (Tomlinson et al., 2023). This diverse set of scenarios allows construction companies to explore a wide range of potential outcomes, considering multiple factors and variables that might impact their projects and operations.

Moreover, the cost-effectiveness of AI-generated scenarios is a noteworthy benefit for construction companies. Traditional methods of scenario planning and analysis could be time-consuming and resource-intensive. However, with AI at their disposal, construction companies can generate numerous scenarios at a relatively low cost, saving both time and resources. This efficiency is especially advantageous in the fast-paced and competitive construction industry, where staying ahead of emerging trends and challenges is crucial.

AI-generated scenarios also enable construction companies to be more proactive and agile in their strategic planning. By anticipating various future scenarios, managers can develop contingency plans and risk mitigation strategies. This proactive approach ensures that construction companies are better prepared to navigate uncertainties and capitalize on emerging opportunities, bolstering their overall resilience and adaptability.

However, while AI-generated scenarios hold great potential, it is crucial for construction companies to approach their use with caution. Ensuring the quality and accuracy of AI-generated scenarios is paramount to derive meaningful insights. Companies must carefully validate the data and algorithms used to generate these scenarios and be aware of potential biases or limitations in the AI models.

In conclusion, AI-generated scenarios present a valuable tool for construction companies in strategic planning. By leveraging AI's capabilities to explore a wide range of possibilities and anticipate future challenges, construction companies can make informed decisions, optimize their resource allocation, and enhance their overall competitiveness. With careful consideration and validation, AI-generated scenarios can serve as a pivotal asset for construction companies, helping them navigate a dynamic and evolving industry landscape with greater confidence and foresight.

Challenges and Considerations

Indeed, while AI-generated scenarios can provide valuable insights and benefits, it is essential to be mindful of the associated challenges and considerations. Ethical and organizational aspects should be carefully addressed to ensure responsible and effective use of AI-generated scenarios in strategic planning. One critical consideration is the ethical implications of AI-generated scenarios. As AI models rely on vast amounts of data to generate scenarios, ensuring data privacy and security is paramount. Construction companies must safeguard sensitive information and adhere to data protection regulations to protect the privacy of individuals and stakeholders involved.

Additionally, it is crucial to be vigilant about potential biases in AI-generated scenarios. AI models are trained on historical data, and if this data contains biases, it may lead to biased scenarios. Managers must actively monitor and evaluate AI algorithms to identify and rectify any biases that might arise.

Moreover, recognizing the limitations of AI technology is vital. AI-generated scenarios are based on historical data and patterns, which means they may not fully capture entirely novel or unprecedented events. Human expertise and judgment remain essential in complementing AI-generated scenarios to account for unique and unpredictable situations. Organizational readiness is another key consideration. Integrating AI-generated scenarios into the strategic planning process may require organizational adjustments, including the creation of new procedures, skill development, and training for staff. Ensuring that all relevant stakeholders are on board and prepared for these changes is crucial for a smooth and successful implementation.

To address these challenges, construction companies should adopt a proactive and transparent approach to using AI-generated scenarios. Regular auditing and validation of AI models can help identify and correct biases or limitations. Creating guidelines and protocols for responsible AI usage can ensure that ethical considerations are embedded in the decision-making process.

Moreover, fostering a culture of continuous learning and development within the organization will aid in building the necessary skills and capabilities (Saad, Alaloul, Ammad et al., 2022) to effectively leverage AI-generated scenarios. Collaborating with AI experts and ethical consultants can also provide valuable insights and guidance in navigating the complexities of AI in strategic planning.

In conclusion, while AI-generated scenarios offer significant benefits in strategic planning for construction companies, it is essential to acknowledge and address the ethical and organizational considerations associated with their use. By promoting responsible AI practices, ensuring data privacy, and being aware of biases and limitations, construction companies can harness the power of AI-generated scenarios to make informed decisions and drive success in a rapidly evolving industry landscape.

AI-assisted Scenario Development

AI can play a valuable role in scenario development for construction companies by providing foundational material and stimulating discussions with clients. Although AI may not entirely replace human scenarists, its assistance can prove highly beneficial in eliciting and refining client requirements and uncertainties.

AI-assisted scenario development offers significant promise as it can efficiently analyze vast amounts of data, identify patterns, and generate a wide range of potential scenarios. By presenting these scenarios to clients, AI can spark meaningful discussions, encouraging clients to articulate their specific needs and uncertainties more effectively. This interaction helps construction companies gain deeper insights into client preferences, priorities, and risk perceptions, thereby enhancing the overall quality and relevance of the scenarios developed.

Moreover, AI-generated scenarios act as a valuable starting point for human scenarists, allowing them to build upon the generated material and further refine the scenarios. This collaboration between AI and human experts facilitates a synergistic approach to scenario development, harnessing the strengths of both AI-driven data analysis and human creativity and expertise.

The Importance of a Multistage Approach

For creating reliable scenarios, a multi-stage process is essential. This method entails a number of phases, including defining the user and the scenarios' goals, choosing the best selection of scenarios, and adjusting them to the client's requirements. Organizations may make sure that the scenarios they employ to support their strategic planning process are pertinent, accurate, and successful by adopting this method.

Clarifying the User and Purpose

It is essential to clarify who the scenarios are for and the intended purpose of using them. This helps align the scenarios with the specific needs and goals of the organization. By identifying the user and purpose upfront, organizations can ensure that the scenarios generated by AI are tailored to their requirements and can provide valuable insights for decision-making.

Identifying the most Appropriate Scenarios

Not all scenarios are equally relevant or useful for every organization. It is crucial to select the most appropriate set of scenarios that align with the uncertainties and challenges faced by the organization. This selection process involves considering factors such as the significance of uncertainties and the potential impact of different scenarios on the organization's strategy.

Tailoring Scenarios to the Needs of the Client

The scenarios should be adjusted to the client's particular requirements once they have been discovered. This entails figuring out what elements might be missing from each scenario and improving those elements. By customizing the scenarios, organizations can ensure that they address their unique challenges and provide actionable insights for strategy development.

Ethical Concerns and Considerations Associated with the Use of AI-generated Scenarios

Bias and Discrimination

AI-generated situations could lead to ethical issues with prejudice and discrimination. Users should be aware of these issues and take precautions to address them, such as routinely evaluating and updating the information and algorithms used to produce the scenarios.

Organizational Changes

Incorporating AI-generated scenarios into the strategic planning process may require organizational adjustments, including the establishment of new procedures and acquisition of specific skills. Users must be prepared to allocate resources, both in terms of time and finances, to effectively implement these changes and ensure buy-in from all relevant stakeholders. This commitment is crucial to successfully integrate AI-generated scenarios and derive maximum value from their application in strategic planning endeavors.

Integration into Strategic Planning

AI-generated scenarios should be integrated into the broader strategic planning process and not used in isolation. Users should consider how these scenarios fit into their overall strategy and how they can be used to support decision making and action planning.

Dark Forecasting

Dark forecasting involves utilizing machine learning and data-driven methodologies to predict undesirable events, such as crimes, accidents, or

Table 8.2: Key findings, benefits, challenges, and best practices for incorporating AI-generated scenarios in strategic planning

Key findings and benefits	Challenges and considerations
AI-generated scenarios provide valuable insights to managers, enabling better decision-making.	Ethical implications of bias and discrimination in the underlying data and algorithms. Organizational adjustments required for effective integration of AI-generated scenarios.
They enhance the strategic planning process and expand strategic options for companies.	Financial investment for AI technologies and data infrastructure.
Proactive approach with AI-generated scenarios fostersadaptability and resilience. Best Practices.	Collaboration and buy-in from stakeholders for successful implementation of AI-assisted Scenario Development.
Increase "Futures Literacy" of employees through regular discussions and analysis of AI-generated scenarios.	AI can efficiently analyze data, identify patterns, and generate various scenarios to stimulate meaningful client interactions.
Address ethical implications and potential biases in AI models and data.	AI-generated scenarios serve as foundational material for human scenarists to build upon and refine scenarios.

terrorist attacks. Although this approach can be valuable for prevention and response purposes, it is worth noting that the term "dark forecasting" may carry negative associations. It is important to consider the ethical implications and potential biases associated with such predictions.

Table 8.2 states that integration of AI-generated scenarios into various processes offers significant opportunities; however, it also brings forth ethical challenges. Users must remain cognizant of these issues and implement measures to promote fairness, transparency, and accountability in the utilization of AI technologies.

Case Studies of AI in Fabrication and Construction

The integration of AI technologies into practical applications has emerged as a transformative paradigm within the fabrication and construction industries, offering profound prospects for augmenting efficiency, precision, and productivity. This pioneering utilization of AI encompasses a wide array of innovative tools and methodologies, such as generative design, predictive analytics, augmented reality, and computer vision, which collectively empower these industries to optimize their operational processes and streamline various facets of project execution (Walmsley, 2022). Through the judicious application of AI-driven solutions, real-world scenarios have demonstrated the potential to revolutionize conventional practices, ushering in a new era of enhanced performance and efficacy across the fabrication and construction domains. As AI continues to evolve, its manifold benefits hold the promise of revolutionizing traditional methods, elevating the industries to unparalleled heights of competitiveness and sustainability (Oztemel & Gursev, 2020).

Autodesk's Generative Design for AEC

To benefit the Architecture, Engineering, and Construction (AEC) business, generative design technology has made great progress thanks in a large part to Autodesk, a well-known software provider. This cutting-edge technology makes use of sophisticated AI algorithms to explore a wide range of design choices while abiding by predetermined limitations and goals. One noteworthy case study exemplifying the integration of generative design is showcased in the Autodesk MaRS office project in Toronto, Canada. The application of generative design tools facilitated the optimization of the office building's layout, resulting in a highly efficient and sustainable design solution.

Through the iterative exploration of countless design possibilities, generative design enabled architects to arrive at an optimal layout that not only met functional requirements but also exhibited exceptional resource utilization. By significantly reducing material waste and streamlining the construction timeline, the project demonstrated the remarkable potential of AI to revolutionize conventional architectural design and construction practices.

This case study serves as an illustration of how AI-driven generative design could assist the AEC sector in constructing a more resource-conserving and sustainable future. How structures are envisaged, planned, and constructed may be completely changed by the use of AI technology in architectural design processes. By doing so, the sector will be moved toward becoming more efficient and sustainable, pushing the boundaries of innovation. Architects and construction experts now have a previously unheard-of potential to design more sustainable and optimal built environments for the future as a result of AI's integration into practical applications as it continues to develop and mature.

AI-powered Prefabrication at Katerra

Katerra, a pioneering company in construction technology, has embarked on a transformative journey by integrating AI and robotics into the prefabrication process. By strategically applying AI algorithms for design and planning, Katerra has successfully streamlined the production of building components within its cutting-edge factories. The seamless integration of robotics and AI in the prefabrication process has yielded remarkable outcomes, characterized by heightened precision, reduced labor costs, and accelerated construction timelines (De Valence, 2018).

Driven by a commitment to enhanced efficiency and innovation, Katerra has undertaken a noteworthy case study involving the construction of a 7-storey hotel in Spokane, Washington. Leveraging their advanced prefabrication techniques, the company achieved significant reductions in construction time, setting new benchmarks for the industry. The success of this case study stands as a compelling testament to the potential of AI and robotics in revolutionizing conventional construction practices (De Valence, 2018).

The strategic implementation of AI and robotics in the prefabrication process has established Katerra as a trailblazer in the industry, effectively demonstrating the profound impact of technology on construction. Katerra has demonstrated the potential for attaining accuracy, cost-effectiveness, and expediency in the production of prefabricated building components by carefully utilizing AI algorithms to optimize design and planning and the seamless integration of robotics into the production line (De Valence, 2018).

The extraordinary integration of robotics and AI in the construction process portends a bright future for the sector, in which cutting-edge technologies will fundamentally alter conventional construction techniques. The Spokane, Washington, case study success by Katerra acts as a source of motivation for the construction industry, inspiring other stakeholders to investigate and embrace the revolutionary potential of AI and robots in creating sustainable and effective construction projects (De Valence, 2018).

Predictive Analytics for Construction Equipment Maintenance

In the construction industry, the integration of AI and predictive analytics has facilitated the optimization of equipment maintenance schedules. Prominent companies, such as Caterpillar, have successfully implemented AI-based systems that analyze sensor data from heavy machinery to predict and identify potential breakdowns and maintenance requirements (Bayzid et al., 2016). By harnessing the power of predictive analytics, construction firms can proactively schedule maintenance tasks, thereby minimizing downtime and enhancing the reliability of equipment. This innovative approach has demonstrated significant cost savings and remarkable improvements in overall project efficiency (Bayzid et al., 2016).

Augmented Reality (AR) in Construction

Since it provides workers on the job site with access to real-time information and visualization tools, the usage of augmented reality (AR) in the construction industry has proven to be invaluable (Loporcaro et al., 2020).

The Trimble Connect platform, which seamlessly merges AR and BIM technologies, was presented by renowned technology company Trimble (Loporcaro et al., 2020).

AI-driven Quality Control in Steel Fabrication

The incorporation of AI technologies has extended to quality control in steel fabrication processes, contributing to notable advancements in the industry. Esteemed companies like Civimec, an Australian fabrication company, have embraced AI-based computer vision systems to enhance quality control measures by detecting defects and deviations in steel components (Chadha et al., 2022). The utilization of these AI-driven quality control systems assures the accuracy and precision of fabricated steel parts, ultimately fostering heightened safety and reliability in construction projects. The application of AI in steel fabrication serves as a pivotal enabler for achieving superior construction outcomes and reinforces the significance of technological integration in modern-day fabrication practices (Chadha et al., 2022).

In conclusion the case studies presented herein exemplify the myriad applications of AI within the realms of fabrication and construction, serving as compelling evidence of its transformative capabilities across various facets of these industries. By delving into areas such as design optimization, prefabrication, predictive maintenance, and quality control, AI technologies manifest as promising solutions that hold the potential to elevate efficiency, sustainability, and safety within construction processes. These pioneering implementations underscore the ever-increasing significance of AI in reshaping conventional practices and propelling the industries towards a more technologically empowered and progressive future.

AI in Fabrication and Construction: Prefabrication and Modularization

The emergence of AI has completely transformed a number of industries, and the fabrication and construction industry is no exception. Prefabrication and modularization procedures that incorporate AI technology have become a cutting-edge trend, giving several benefits in terms of effectiveness, affordability, and sustainability. Prefabrication and modularization have long been used in the building industry to speed up construction and increase production. With the infusion of AI, these methodologies are now undergoing a transformative shift, paving the way for more innovative and dynamic approaches to building construction (Sun, 2021).

Prefabrication refers to the manufacturing of building components off-site in a controlled environment, followed by their assembly on the construction site. Higher precision, shorter construction times, and better-quality control are all possible with this technique. On the other hand, modularization entails the development of independent building pieces or modules that can be joined to make bigger buildings. Due to their capacity to shorten on-site disturbances and speed up building timetables, both techniques have grown in favor (Sun, 2021).

AI-driven technologies have the potential to revolutionize the fabrication and construction processes in prefabrication and modularization. By leveraging ML algorithms, computer vision, and advanced robotics, AI can optimize various stages of these methodologies, from design and material selection to manufacturing and assembly. AI's predictive capabilities enable early detection of design flaws, leading to enhanced structural integrity and energy efficiency (Ahmad et al., 2021). Furthermore, AI-driven automation and robotics streamline the fabrication process, reducing human errors and labor costs.

The application of AI in prefabrication and modularization also has far-reaching implications for sustainability and resource optimization. By analyzing vast datasets and historical performance data, AI can assist in the selection of environmentally friendly materials, minimize waste, and optimize energy consumption during the manufacturing and assembly stages (Debrah et al., 2022). This convergence of AI and prefabrication/modularization aligns perfectly with the global push for eco-friendly and sustainable construction practices.

However, the integration of AI into fabrication and construction processes also presents unique challenges and considerations. Ethical implications regarding data privacy, transparency, and bias need to be addressed to ensure responsible AI deployment. Moreover, the adoption of AI requires a shift in the traditional construction industry mindset, necessitating workforce upskilling and training to effectively utilize AI-driven technologies.

AI in Fabrication and Construction: On-site Assembly and Installation

In recent years, the construction and fabrication industries have undergone a significant transformation due to the escalating integration of AI technologies. AI has emerged as a pivotal driver of innovation, offering novel avenues for on-site assembly and installation processes within the realms of fabrication and construction. By harnessing AI-driven methodologies, these industries are experiencing noteworthy enhancements in efficiency, accuracy, and safety, thereby propelling them toward a more sustainable and technologically advanced future (Jiang et al., 2022).

The assimilation of AI in on-site assembly and installation endeavors addresses several challenges confronted by conventional construction practices. These challenges encompass labor-intensive procedures, potential human errors, and limitations in precision. Leveraging its capacity to analyze copious datasets, learn from patterns, and make informed decisions, AI has presented unparalleled opportunities for process optimization, cost reduction, and augmented productivity.

Moreover, the ethical implications associated with AI's adoption within the construction domain warrant considerable attention. As AI increasingly becomes an integral component of daily practices, concerns pertaining to data privacy, bias mitigation, and human-robot collaboration take center stage.

The potential impact of AI on the fabrication and construction industries is of profound significance, as it holds the promise of transforming age-old practices and paving the way for intelligent, data-driven, and autonomous construction processes. By scrutinizing the prevailing state of AI adoption, identifying existing challenges, and envisioning future possibilities (Jiang et al., 2022).

The construction industry is witnessing a paradigm shift with the increasing adoption of modular integrated construction (MIC), a revolutionary approach that offers enhanced quality, productivity, and sustainability. MIC transforms traditional cast-in-situ processes into on-site assembly of prefabricated modules, presenting a game-changing potential in construction practices (Wuni & Shen, 2020). Particularly, the on-site assembly stage in MIC scenarios is both uncertain and complex, given the variability of outside conditions, multi-contractor organization, and geographic dispersion of activities.

To address the challenges posed during on-site assembly, the integration of Information and Communication Technology (ICT) has been pursued. However, the full digitalization and cyber-physical interoperation of on-site resources have faced obstacles, leading to fragmented and out-of-date cyber-physical coordination (Saad, Alaloul, Rasheed et al., 2022). Digital twin, as a comprehensive digital representation of entities or activities, emerges as a pivotal enabler in the ICT revolution for real-time convergence and coordination.

AI in Fabrication and Construction: Smart Sensing and Monitoring

AI is important in the manufacturing and construction industries because it enables smart sensing and monitoring. Sensors with AI integration can collect real-time data and evaluate it to make smart judgments. This technology helps in optimizing processes, improving efficiency, and ensuring safety in construction projects. AI-powered sensors can detect and predict potential issues, monitor equipment performance, and enhance overall productivity (Rasheed et al., 2021). By leveraging AI in fabrication and construction, companies can achieve smarter and more sustainable practices (Chaudhary et al., 2022).

The integration of AI with smart sensing and monitoring systems empowers them to go beyond mere data collection, transforming them into intelligent and adaptive entities. AI algorithms, particularly machine learning techniques, allow these systems to analyze and learn from the collected data, identifying patterns, anomalies, and trends that might otherwise go unnoticed. This ability to recognize complex patterns and relationships enhances the accuracy and predictive capabilities of these systems, making them invaluable in various industries and sectors (Baduge et al., 2022).

One of the primary advantages of AI-driven smart sensing and monitoring lies in its ability to enable proactive decision-making and predictive maintenance. By continuously analyzing real-time data, these systems can identify potential issues, anomalies, or deviations from normal operating conditions. This early detection allows for timely interventions, minimizing the risk of system failures, accidents, or costly downtime. Moreover, the data-driven insights gained from these systems can inform better resource management, leading to more efficient utilization and reduced waste.

The potential applications of AI in smart sensing and monitoring are vast and diverse. In the realm of smart cities, AI-powered monitoring systems can optimize traffic flow, enhance public safety, and efficiently manage energy consumption. In healthcare, AI-driven sensors can revolutionize patient monitoring, enabling remote health tracking, personalized treatment plans, and early detection of health issues (Naeem et al., 2022). Industries like manufacturing, agriculture, and logistics can leverage AI-driven sensing and monitoring for process optimization, quality control, and supply chain management.

However, as with any technology, the integration of AI in smart sensing and monitoring also presents certain challenges and ethical considerations. Privacy and data security are critical concerns, especially when dealing with sensitive data in healthcare or personal information in smart city applications. To avoid biased decision-making or discriminatory outcomes, AI systems must be fair and transparent. Furthermore, the seamless integration of AI with existing systems, as well as the interoperability of various sensors and devices, are technological obstacles that must be overcome before widespread deployment (Amiri-Zarandi et al., 2022).

References

Abbasi, M.F., Bilal, M. and Rasheed, K. (2022). Role of Human Intuition in AI Aided Managerial Decision Making: A Review. *2022 International Conference on Decision Aid Sciences and Applications, DASA 2022*. https://doi.org/10.1109/DASA54658.2022.9765153.

Abdul Hannan Qureshi, Wesam Salah Alaloul, Manzoor, B., Musarat, M.A., Saad, S. and Ammad, S. (2020). Implications of Machine Learning Integrated Technologies for Construction Progress Detection under Industry 4.0 (IR 4.0). *2020 Second International Sustainability and Resilience Conference: Technology and Innovation in Building Designs, 0*. https://doi.org/10.1109/IEEECONF51154.2020.9319974.

Ahmad, T., Zhang, D., Huang, C., Zhang, H., Dai, N., Song, Y. and Chen, H. (2021). Artificial intelligence in sustainable energy industry: Status Quo, challenges and opportunities. *Journal of Cleaner Production*, 289, 125834.

Altaf, M., Alaloul, W.S., Musarat, M.A., Bukhari, H., Saad, S. and Syed Ammad. (2020). BIM Implication of Life Cycle Cost Analysis in Construction Project: A Systematic Review. *2020 Second International Sustainability and Resilience Conference: Technology and Innovation in Building Designs*. https://doi.org/10.1109/IEEECONF51154.2020.9319970.

Amiri-Zarandi, M., Hazrati Fard, M., Yousefinaghani, S., Kaviani, M. and Dara, R. (2022). A platform approach to smart farm information processing. *Agriculture*, 12(6). https://doi.org/10.3390/agriculture12060838.

Ammad, S., Alaloul, W.S., Saad, S., Altaf, M., Alawag, A.M. and Ali, M. (2021). Building Information Modeling (BIM) and Occupational Safety in Infrastructure Projects. *2021 International Conference on Data Analytics for Business and Industry (ICDABI)*, pp. 240–244. https://doi.org/10.1109/ICDABI53623.2021.9655832.

Ammad, S., Saad, S., Bashir, M.T., Qureshi, A.H., Altaf, M. and Rasheed, K. (2021). Construction Accidents via Integrating Building Information Modeling (BIM) with Emerging Digital Technologies: A Review. *2021 3rd International Sustainability and Resilience Conference: Climate Change*. https://doi.org/10.1109/IEEECONF53624.2021.9668066.

Author, C., Danesh Narooei, K. and Ramli, R. (2014). Application of artificial intelligence methods of tool path optimization in CNC machines: A review. *Research Journal of Applied Sciences, Engineering and Technology*, 8(6), 746–754.

Baduge, S.K., Thilakarathna, S., Perera, J.S., Arashpour, M., Sharafi, P., Teodosio, B., Shringi, A. and Mendis, P. (2022). Artificial intelligence and smart vision for building and construction 4.0: Machine and deep learning methods and applications. *Automation in Construction*, 141, 104440. https://doi.org/https://doi.org/10.1016/j.autcon.2022.104440.

Bahroun, Z., Anane, C., Ahmed, V. and Zacca, A. (2023). Transforming education: A comprehensive review of generative artificial intelligence in educational settings through bibliometric and content analysis. *Sustainability*, 15(17). https://doi.org/10.3390/su151712983.

Bayzid, S.M., Mohamed, Y. and Al-Hussein, M. (2016). Prediction of maintenance cost for road construction equipment: A case study. 43(5), 480–492. https://doi.org/10.1139/CJCE-2014-0500.

Chadha, U., Selvaraj, S.K., Raj, A., Mahanth, T., Praveen Vignesh, S.T., Lakshmi, P.J., Samhitha, K., Reddy, N.B. and Adefris, A. (2022). AI-driven techniques for controlling the metal melting production: A review, processes, enabling technologies, solutions, and research challenges. *Materials Research Express*, 9(7), 072001. https://doi.org/10.1088/2053-1591/AC7B70.

Chaudhary, V., Kaushik, A., Furukawa, H. and Khosla, A. (2022). Review: Towards 5th generation AI and IoT driven sustainable intelligent sensors based on 2D MXenes and Borophene. *ECS Sensors Plus*, 1(1), 013601. https://doi.org/10.1149/2754-2726/ac5ac6.

De Valence, G. (2018). Three Pathways to the Future for Building and Construction. *First International Conference on Construction Futures*. Wolverhampton.

Debrah, C., Chan, A.P.C. and Darko, A. (2022). Artificial intelligence in green building. *Automation in Construction*, 137, 104192. https://doi.org/https://doi.org/10.1016/j.autcon.2022.104192.

Gil, D., Hobson, S., Mojsilović, A., Puri, R. and Smith, J.R. (2020). AI for management: An overview. *In:* J. Canals and F. Heukamp (Eds.), *The Future of Management in an AI World: Redefining Purpose and Strategy in the Fourth Industrial Revolution*, pp. 3–19. Springer International Publishing. https://doi.org/10.1007/978-3-030-20680-2_1.

Hannan Qureshi, A., Salah Alaloul, W., Kai Wing, W., Saad, S., Ali Musarat, M., Ammad, S. and Farouk Kineber, A. (2023). Automated progress monitoring technological model for construction projects. *Ain Shams Engineering Journal*, 14(10), 102165. https://doi.org/https://doi.org/10.1016/j.asej.2023.102165.

Harfoush, A., Haapala, K.R. and Tabei, A. (2021). Application of artificial intelligence in incremental sheet metal forming: A review. *Procedia Manufacturing*, 53, 606–617. https://doi.org/10.1016/J.PROMFG.2021.06.061.

Jiang, Y., Li, M., Guo, D., Wu, W., Zhong, R.Y. and Huang, G.Q. (2022). Digital twin-enabled smart modular integrated construction system for on-site assembly. *Computers in Industry*, 136, 103594. https://doi.org/10.1016/J.COMPIND.2021.103594.

Loporcaro, G., Bellamy, L., McKenzie, P. and Riley, H. (2020). Evaluation of Microsoft HoloLens Augmented Reality Technology as a Construction Checking Tool. *19th International Conference on Construction Applications of Virtual Reality*.

Naeem, U.H., Bilal, M., Syed, F., Rasheed, K. and Saad, S. (2022). A Multilayer Encryption Model to Protect Healthcare Data in Cloud Environment. *2022 International Conference on Data Analytics for Business and Industry, ICDABI 2022*. https://doi.org/10.1109/ICDABI56818.2022.10041708.

Norori, N., Hu, Q., Aellen, F.M., Faraci, F.D. and Tzovara, A. (2021). Addressing bias in big data and AI for health care: A call for open science. *Patterns*, 2(10), 100347. https://doi.org/10.1016/j.patter.2021.100347.

Oztemel, E. and Gursev, S. (2020). Literature review of Industry 4.0 and related technologies. *Journal of Intelligent Manufacturing*, 31(1), 127–182. https://doi.org/10.1007/S10845-018-1433-8/FIGURES/11.

Paschen, U., Pitt, C. and Kietzmann, J. (2020). Artificial intelligence: Building blocks and an innovation typology. *Business Horizons*, 63(2), 147–155. https://doi.org/https://doi.org/10.1016/j.bushor.2019.10.004.

Patil, S. and Shankar, H. (2023). Multidisciplinary sciences and arts transforming healthcare: Harnessing the power of AI in the modern era. *International Journal of Multidisciplinary Sciences and Arts*, 60–70. https://doi.org/10.47709/ijmdsa.v2i1.2513.

Rasheed, K., Shahzad, L., Saad, S., Khan, H.A., Ahmed, W. and Sadiq, T. (2021). Parking Guidance System Using Wireless Sensor Networks. *2021 International Conference on Decision Aid Sciences and Application, DASA 2021*. https://doi.org/10.1109/DASA53625.2021.9682213.

Riege, A. and Lindsay, N. (2006). Knowledge management in the public sector: Stakeholder partnerships in the public policy development. *Journal of Knowledge Management*, 10(3), 24–39. https://doi.org/10.1108/13673270610670830.

Saad, S., Alaloul, W.S., Ammad, S. and Qureshi, A.H. (2022). A qualitative conceptual framework to tackle skill shortages in offsite construction industry: A scientometric approach. *Engineering, Construction and Architectural Management*, 29(10), 3917–3947. https://doi.org/10.1108/ECAM-04-2021-0287.

Saad, S., Alaloul, W.S., Rasheed, K. and Ammad, S. (2022). Modeling and simulation of construction cyber-physical systems. *In: Cyber-Physical Systems in the Construction Sector*, pp. 88–110. CRC Press.

Spaniol, M.J. and Rowland, N.J. (2023). AI-assisted scenario generation for strategic planning. *Futures and Foresight Science*, 5(2), 1–10. https://doi.org/10.1002/ffo2.148.

Sun, J. (2021). Intelligent Construction of Prefabricated Buildings Based on Building Information Modeling and Artificial Intelligence & Internet of Things. *2021 3rd International Academic Exchange Conference on Science and Technology Innovation, IAECST 2021*, 2033–2036. https://doi.org/10.1109/IAECST54258.2021.9695861.

Tomlinson, B., Torrance, A.W. and Black, R.W. (2023). ChatGPT and works scholarly: Best practices and legal pitfalls in writing with AI. *ArXiv Preprint ArXiv:2305.03722*. https://doi.org/10.48550/arXiv.2305.03722

Walmsley, K. (2022). Cutting-edge practical research on generative design, IoT and digital twins. *Innovation in Construction*, 117–145. https://doi.org/10.1007/978-3-030-95798-8_7/COVER.

Wang, Y. Bin, Zheng, P., Peng, T., Yang, H.Y. and Zou, J. (2020). Smart additive manufacturing: Current artificial intelligence-enabled methods and future perspectives. *Science China Technological Sciences*, 63(9), 1600–1611. https://doi.org/10.1007/s11431-020-1581-2.

Williams, R. and Yampolskiy, R. (2021). Understanding and avoiding AI failures: A practical guide. *Philosophies*, 6(3). https://doi.org/10.3390/philosophies6030053.

Wuni, I.Y. and Shen, G.Q. (2020). Fuzzy modeling of the critical failure factors for modular integrated construction projects. *Journal of Cleaner Production*, 264, 121595. https://doi.org/https://doi.org/10.1016/j.jclepro.2020.121595.

Zanoni, S., Ashourpour, M., Bacchetti, A., Zanardini, M. and Perona, M. (2019). Supply chain implications of additive manufacturing: A holistic synopsis through a collection of case studies. *International Journal of Advanced Manufacturing Technology*, 102(9–12), 3325–3340. https://doi.org/10.1007/S00170-019-03430-W/METRICS.

CHAPTER

9

Smart Infrastructure and AI

Muhammad Waqas Khan, Syed Saad, Syed Ammad, Kumeel Rasheed and Qaiser Jamal

Introduction

As humanity steps up boldly into the 21st century, one truth has become abundantly clear that our world is growing more linked, complicated, and needing creative solutions to address its most critical concerns. Urbanization and climate change are two current concerns that require more collaboration, efficiency, and creativity than we could have imagined ten years ago. In this setting, technologies like "smart infrastructure" and developments in artificial intelligence (AI) are positioned to alter our way of life fundamentally. Infrastructure, the fundamental physical systems we rely on for transportation, electricity, water, and other purposes, have aged in many ways and is unable to keep up with the modern demands or take advantage of new developments in technology. While this is going on, AI techniques like machine learning (ML) have been developed that can discover hidden patterns in massive amounts of data and even acquire new abilities on their own.

Smart Infrastructure and Its Applications

Modern technologies like AI, the Internet of Things (IoT), and data analytics have changed how we plan, create, and maintain our cities and essential infrastructure systems. This chapter will cover the various uses that smart infrastructure makes possible, demonstrating its revolutionary potential and the wealth of advantages it provides to society. Modern society has undergone a paradigm shift in how we conceptualize, create, and manage our urban centers and vital infrastructure networks as a result of the integration of cutting-edge technologies like AI, IoT, and advanced data analytics. Thereafter, the chapter will focus on this fundamental transformation and explore the wide range of applications that smart infrastructure enables, emphasizing both its profound ability to upend social norms and its broad range of positive effects on society as a whole. The chapter acts as a conduit to study the dynamic interaction between these cutting-edge technologies and the structural underpinnings of our cities and essential infrastructure systems. This investigation includes a thorough investigation of how cognitive AI, IoT, and the complex field of data analytics intersect to transform urban planning, building methodology, and operational maintenance tactics. A new age of possibilities is opened up by the seamless fusion of these technological forces, demonstrating their synergistic power to surpass traditional boundaries.

Smart Transportation

Transportation comes to be the initial area of attention. The city's highways now include smart sensors and traffic control systems driven by AI, allowing for real-time monitoring and optimization of traffic flow (Badshah et al., 2020). As a result, travel times decreased, congestion in traffic was much reduced, and fuel usage fell. Smart applications that gave real-time updates on traffic conditions, public transit timetables, and availability of parking allowed commuters to plan their routes, resulting in a smooth and effective travel experience (Xiong et al., 2012). The initial phase's main area of concentration is the field of transportation, which has become increasingly important. Urban thoroughfares in this area have experienced a significant makeover, combining sophisticated sensors and traffic control systems that depend on AI for their operation. This connection represents a paradigm leap, enabling cities to monitor traffic patterns in real-time and optimize them precisely. This significant change has had observable advantages, with the commuters' trip times being among the most notable. This clever orchestration of traffic dynamics has greatly reduced the prevalent problem of congestion, a long-standing problem of urban centers, leading to a noticeable improvement in the flow of vehicle travel (Bhogaraju & Korupalli, 2020). The significant decrease in fuel consumption that results from this integration is a noteworthy accomplishment brought about by the improved effectiveness of traffic management. In addition to helping residents save money, this reduction supports the city's larger sustainability objectives and is in line with the need to reduce carbon emissions and environmental effect (Bhogaraju & Korupalli, 2020). Importantly, this intelligent infrastructure's novel features go beyond simple traffic optimization. Real-time insights into a wide range of crucial data are now available thanks to a tapestry of intelligent apps. The opportunity to get real-time information on traffic conditions, keep up with the nuances of public transport schedules, and check the availability of parking spots has been granted to commuters (Wang et al., 2006).

Smart Energy Management

To maximize the usage of electricity and cut down on carbon emissions, the city also puts in place sophisticated energy management systems (Saad, Alaloul et al., 2020). Energy consumption trends, weather predictions, and real-time data from smart meters were all examined by AI algorithms to estimate demand and modify electricity generation appropriately. Through the use of smart home technologies, residents were given the ability to monitor and manage their energy use, which decreased energy costs and improved the sustainability of the environment (O'Dwyer et al., 2019).

AI in Smart Infrastructure Design and Maintenance

The use of AI into a variety of sectors has recently altered several industries and the way we live and work. Designing and maintaining smart infrastructure is one area where AI is making major advancements. AI contributes to the field of smart infrastructure, displaying its potential to better maintenance procedures, optimize design processes, and increase overall effectiveness and sustainability (Huseien & Shah, 2022). A rebellious revolution has recently been launched by the widespread use of AI across several industries, redefining whole sectors and radically changing the way that we live and work. One particularly notable aspect of this broad technical progression is the introduction of AI to the field of smart infrastructure. The way we design, build, and maintain our built environment is being transformed by this infusion, which not only has enormous promise but is already making significant progress in that direction (Nasr et al., 2020). Foremost, the infusion of AI-driven solutions into smart infrastructure heralds a new era of maintenance practices. Traditional maintenance approaches have often been reactive, responding to issues only after they arise. In stark contrast, AI empowers predictive upkeep by harnessing the prowess of data analytics and ML algorithms. These tools analyze real-time data streams from sensors embedded within the infrastructure, enabling the early detection of anomalies and degradation. By doing so, AI not only minimizes downtime but also averts catastrophic failures, thereby not just extending the lifecycle of these assets but also fortifying the safety of the communities relying on them (Mehmood et al., 2020) .

AI-driven Design Optimization

The use of AI in the design of smart infrastructure has the potential to accelerate and improve the entire procedure. AI can analyze a large quantity of data from many sources, including environmental conditions, human behavior, and historical trends. With the help of this study, engineers and architects may develop more effective and environmentally friendly infrastructure designs. AI algorithms, for instance, may optimize the design of a smart city to reduce traffic jams, increase energy efficiency, and improve public safety (Şerban and Lytras, 2020). A whole new age of transformational possibilities, set to speed up and improve the entire procedural environment, has begun with the incorporation of AI into the design of smart infrastructure. In this rapidly changing world, AI emerges as a powerful orchestrator capable of deftly analyzing enormous data sets culled from several sources, including environmental factors, behavioral patterns of humans, and historical trends. This informational synthesis spurs creativity, enabling engineers and architects to create infrastructure solutions that not only boost productivity but also encourage environmental stewardship (Adochiei et al., 2020). Moreover, under the expert strokes of paint of AI, the canvas of energy efficiency unfolds. AI analyses consumption trends, peaks, and valleys, and optimizes the distribution of energy resources. The energy requirements of the infrastructure are matched with present demands in this choreography of data and algorithms, which results in a seamless interaction between supply and consumption. A tangible decrease in carbon footprints is the result, demonstrating AI's ability to be a leader in sustainable development (Hoosain et al., 2020). The effects of AI go well beyond the visible components as the fabric of smart infrastructures' architecture is still being spun. It penetrates the very core of urban life, sculpting public areas that reflect the goals of the neighborhood. Today, parks, plazas, and recreational spaces are sanctuary-like examples of deliberate design, rather than accidental additions.

Predictive Maintenance and Asset Management

Infrastructure durability and good function depend on regular maintenance. By analyzing real-time data from sensors integrated into buildings and systems, AI plays a crucial part in predictive maintenance. AI systems can identify probable problems or breakdowns before they happen by monitoring variables like temperature, vibration, and stress levels. With the help of this proactive strategy, maintenance personnel can plan repairs and replacements in advance, cutting downtime and preventing expensive emergencies (Liu & Zhang, 2020). An interesting example of this would be a smart bridge that uses AI to monitor structural integrity, foresee repair requirements, and avert disasters in the future (Mehmood et al., 2020). The story of a smart bridge vividly illustrates an astounding expression of AI's predictive power. This technological marvel serves not only as a means of transportation but also as proof of AI's sentinel function. Here, AI is keeping a close check on the structural stability of the bridge down to its very fibers. The analytical crucible of AI is filled with data points that are as varied as the heartbeat of the bridge, ranging from temperature changes to the frequency of vibrations. As patterns emerge, AI unleashes its prophetic skills, anticipating upcoming corrosion, wear and tear, or weaknesses that could otherwise go unnoticed by regular human inspection. The maintenance story changes in this complex dance between data and algorithms (S. Ahmed et al., 2020).

Energy Optimization and Sustainability

AI has a significant impact on energy efficiency and sustainability in smart infrastructure. AI algorithms can optimize energy distribution, maximize the use of renewable energy sources, and decrease total consumption by examining data from smart grids, weather predictions, and energy use trends. AI-enabled smart buildings, for instance, may automatically modify lighting, heating, and cooling systems depending on occupancy patterns, resulting in considerable energy savings. It would be interesting and pertinent to tell a tale about how AI-driven energy optimization affects a city's carbon footprint and energy prices. The ability of AI to maximize the use of renewable energy sources is among the narrative's finest achievements. AI algorithms adjust the injection of solar, wind, and hydro-generated electricity into the grid, painstakingly matching it with demand peaks and supply troughs, much like a conductor fine-tuning an orchestra. AI advances cities towards a future where clean energy takes front stage, reducing greenhouse gas emissions and boosting ecological glory (Yigitcanlar et al., 2021). This is accomplished by utilizing previous use trends, forecasting weather patterns, and optimizing energy storage (Mehmood et al., 2020). However, the story of AI's significance flows beyond the world of technology, reverberating in the design of intelligent structures. Here, AI and infrastructure design work together to create intelligent structures that can adjust to the rhythms of people. Imagine a city where intelligent structures are endowed with a sensitive consciousness that can sense changes in occupancy. With the ebb and flow of people in mind, AI algorithms built into these buildings choreograph a smooth dance between the lighting, heating, and cooling systems. Energy is not squandered in dark spaces, temperatures are kept in perfect harmony with the outside conditions, and comfort is meticulously maintained. The end result is an energy efficiency ballet where waste is kept to a minimum and every kilowatt is valued (Mehmood et al., 2020).

Case Studies of Smart Infrastructure and AI

The incorporation AI has led to game-changing breakthroughs and revolutionary undertakings in the quickly changing smart infrastructure market. The breadth of AI applications is extensive as readers will learn a lot about the practical uses, advantages, and difficulties of using AI in the field of smart infrastructure through these real-world examples. Readers will learn a lot about the practical uses, advantages, and difficulties of using AI in the field of smart infrastructure through these real-world examples. The introduction of AI has acted as a game-changing catalyst in the dynamic world of smart infrastructure, sparking innovations that have changed the entire structure of this fast-developing field (Curley & Salmelin, 2018). A new age characterized by innovative endeavors that resound as game-changers within the fabric of modernity has been ushered in by the incorporation of AI. Table 9.1 represents the fusion of infrastructure and technology which demonstrate utilization as well as a range of benefits and difficulties, helping to create a narrative of development and opportunity.

Smart Grid Optimization (The Energy City)

The Energy City is a groundbreaking initiative that uses AI-driven smart grids to

Table 9.1: Exploring the spectrum: Applications of AI in smart infrastructure

Application	Description	Importance	Key differences
Predictive maintenance	Utilizes AI and IoT data to predict equipment breakdowns and optimize maintenance schedules.	Extends equipment life, reduces downtime and maintenance costs.	Focuses on machinery health and performance.
Asset management	AI-driven system for tracking and managing assets efficiently throughout their lifecycle.	Enhanced resource allocation, increased efficiency in asset management.	Broad scope, including tracking and managing various types of assets.
Energy optimization and sustainability	AI analyzes energy consumption patterns and recommends efficiency improvements.	Reduced energy costs, minimized environmental impact.	Concentrates on optimizing energy usage in various sectors for sustainable operation.
Smart grid optimization (The Energy City)	AI manages energy distribution, balances supply and demand, and responds to real-time changes.	Stable energy supply, efficient grid operation.	Focuses on optimizing power distribution and management within a city-scale grid.
Intelligent transportation systems (The Smart Highway)	AI controls traffic flow, reduces congestion, and enhances safety through real-time analysis.	Improved traffic flow, reduced accidents.	Concentrates on enhancing transportation systems through AI and automation.
Smart building automation (The Intelligent Office Tower)	AI-based automation system for building controls such as lighting, HVAC, and security.	Energy efficiency, occupant comfort.	Focuses on optimizing building operations and user experience in commercial buildings.

optimize energy use and advance sustainability. The city's energy infrastructure continually analyses and learns from real-time data, including energy output, demand trends, and weather forecasts, by utilizing cutting-edge machine learning algorithms (Saad et al., 2022). This enables the system to optimize the use of renewable energy sources, minimize waste, and alter the distribution of energy in real time. The Energy City is now positioned as a model for future smart grid installations thanks to its large savings in carbon emissions, cheaper energy prices, and better grid dependability (H. Kim et al., 2021). A cutting-edge project called the Energy City, which is propelled by the seamless integration of AI into the world of smart grids, stands out as a beacon of innovation in the rapidly changing environment. The boundaries of urban growth are redefined by this trailblazing effort, which orchestrates a symphony of energy optimization and sustainability. The Energy City, which is a paradigm shift that combines real-time data analysis with cutting-edge machine learning algorithms, develops as a result of a strong dedication to utilizing the potential of AI (Ahmad et al., 2022). At the center of this innovative project is a complex dance between energy infrastructure and AI, a relationship that puts the city in a class of its own. The city's electrical system serves as a canvas for AI to paint its transforming strokes with an accuracy akin to an artist's brushstroke. Real-time data, gathered from many sources such as demand trends, energy output measures, and weather predictions, converges into a symposium of insights. These observations develop into a dynamic repertory, or a body of information, that the system constantly learns from and adjusts to, as viewed through the lens of ML (H. Kim et al., 2021).

Intelligent Transportation Systems (The Smart Highway)

The Smart Highway is a large-scale initiative that uses AI to improve safety, optimize traffic flow, and lessen congestion on a significant highway network. The Smart Highway gathers real-time data on traffic patterns, weather conditions, and accidents using a networked system of sensors, cameras, and AI algorithms (Ali et al., 2021). After further analysis, this data is used to dynamically modify speed limits, traffic signal timings, and give drivers real-time information via mobile applications and smart signs. The Smart Highway has greatly improved traffic efficiency and commuter comfort by reducing travel time, accidents, and fuel consumption (Bhogaraju & Korupalli, 2020). A groundbreaking project called the Smart Highway emerges as a landmark example of leveraging AI to revolutionize safety, traffic optimization, and congestion relief among the immense expanse of contemporary transportation networks. This enormous project is a monument to innovation, a huge step forward that deftly makes use of AI to alter the characteristics of mobility. The Smart Highway reveals a landscape where technology effortlessly orchestrates the ebb and flow of traffic, raising both efficiency and ease. This is made possible by the convergence of real-time data, a network of sensors and cameras, and the brilliance of AI algorithms (Bhogaraju & Korupalli, 2020).

Smart Building Automation (The Intelligent Office Tower)

The Intelligent Office Tower is a cutting-edge structure that makes use of AI to speed facility management procedures, improve tenant comfort, and optimize energy use. The tower monitors and regulates a number of variables, including lighting, temperature, air quality, and occupancy patterns, by integrating AI-driven building automation technologies. The system learns from occupant preferences using ML algorithms (Qureshi et al., 2020), modifies settings accordingly, and spots chances to increase energy efficiency. The Intelligent Office Tower has established a new benchmark for smart buildings with significant energy savings, lower maintenance costs, and increased tenant satisfaction (Plageras et al., 2018). The Intelligent Office Tower, at the cutting edge of contemporary architecture, is a unique example of innovation, fusing the field of structural design with the capabilities of AI. This innovative facility redefines the idea of smart buildings by establishing a new standard that yields significant benefits in terms of energy reduction, operational effectiveness, and tenant satisfaction (Wang et al., 2006). The tower's AI-powered infrastructure, which was painstakingly built to harmonize a wide range of factors essential to a fluid working environment, is at the heart of its operational excellence. The tower meticulously monitors and fine-tunes various factors in real-time, including lighting intensity, temperature regulation, air quality parameters, and occupancy patterns. This meticulously arranged symphony serves as evidence of the building's dedication to sustainable practices in addition to improving the comfort of its tenants (H. Kim et al., 2021). The Intelligent Office Tower's novel strategy translates into real advantages in a practical sense. It achieves large energy savings that lower maintenance costs while also making a major contribution to environmental goals. A joyful and productive environment is promoted by the harmonic interaction of AI technology and human comfort, which raises tenant satisfaction to new heights (Huseien & Shah, 2022).

Smart Infrastructure and AI: Intelligent Transportation Systems

The way we commute and move regarding in the modern day is changing due to intelligent transport systems and smart infrastructure. These technologies provide game-changing answers to conventional transport networks' problems by utilizing the power of AI. They eventually contribute to the creation of linked and intelligent cities by increasing productivity, enhancing safety, and fostering sustainability. To fully realize the promise of smart infrastructures and AI intelligent transportation systems, we must navigate the future, overcome obstacles, adopt ethical AI practices, and promote collaboration (Xiong et al., 2012). The revolutionary potential of AI underpins the catalytic potential of intelligent transport systems and smart infrastructure. With algorithms that cut through the Gordian knots of traffic jams, productivity snarls, and security flaws, these technologies manifest as game-changers. The intricate choreography of traffic flow is deciphered by AI, which then adapts in real time to optimize routes, foresee and avert accidents, and dynamically regulate traffic lights. This orchestration promises to free urban landscapes from the constraints of congestion in addition to resulting in more efficient travels. However, the story of intelligent transport and infrastructure goes beyond the simple limits of comfort; it unfolds as a roadmap for the development of cities themselves. These technologies' promotion of interconnection creates a city where productivity is increased, safety is woven into daily life, and sustainability is more than a catchphrase—it's a lived reality. Intelligent cities, where the urban sprawl hums to the beat of optimized transport networks, where each node contributes to the grand symphony of development, are ushered in by AI-infused traffic management (Shahid et al., 2021).

The Integration of AI and Infrastructure

Governments and city planners began to look to AI's potential as they saw the need for creative solutions. A network of sensors and smart devices were installed throughout the city's streets, bridges, and public transit systems. These AI-powered devices have the capacity to gather and analyze real-time data from moving objects like cars, people, and weather, among other things. Previously functioning on strict timetables, traffic lights now dynamically modify their timing depending on the volume of traffic. Alex and other commuters like him have quickly transported to their destinations thanks to AI algorithms that optimized routes for public buses and trains. And the public could access all of this data via user-friendly smartphone applications that provided customized suggestions and real-time updates (Xiong et al., 2012). A meticulously constructed network of sensors and smart devices was woven into the very fabric of the city with a goal similar to architects drawing plans for advancement. This digital architecture stretched its grip, producing a complex neural network throbbing with data, taking in everything from the busy streets to the towering bridges and the complicated web of public transit networks. The stage was set for a technological ballet in which AI-powered devices became the virtuoso performers, able to capture and analyze real-time data from a variety of moving objects, including the ups and downs of cars, the rhythmic cadence of pedestrians, the erratic dance of weather patterns, and more (Mehmood et al., 2020). The traffic lights, which were formerly stationary sentinels for traffic control, experienced a metamorphosis in this urban theatre under the guidance of AI. These traffic lights developed into dynamic conductors, their timing fluidly adjusting to the symphony of traffic flow, and were no longer constrained by the rigidity of fixed schedules. The highway was transformed into a stage for AI's dynamic dance, which expertly orchestrated the flow of traffic while also maximizing efficiency and environmental harmony. Through the eyes of people like Alex, who moved about the cityscape with a newfound efficiency and elegance, the story continues. His everyday travels were designed using the identical algorithms that were used to optimize public transit routes (Mehmood et al., 2020).

The Dawn of Intelligent Transportation Systems

Intelligent transport system adoption was not just about efficiency; sustainability was also a factor. The city's carbon footprint decreased as transportation congestion was lessened and idle time was eliminated. As a result of the widespread use of electric and autonomous cars, emissions have decreased even further and the urban environment has changed. People commute changed from being stressful to being easy. They could now catch up on work, read a book, or just unwind throughout the journey. Different types of transportation were synchronized to provide a smooth transition for their travel. What was previously a daily hassle was now a chance for productivity and downtime (Menouar et al., 2017). A new age of efficiency and sustainability-focused change was ushered in by the introduction of intelligent transportation systems. It was more than just a path towards more efficient travel; it was a story that revealed the gift of sustainability, imagining a day when the city's carbon footprint would be reduced, traffic would ease, and idle cars would be a thing of the past. This seamless fusion of ecological care and innovation altered urban dynamics, changing not just how people moved but also the air they breathed (Menouar et al., 2017). This progress is driven by the city's unwavering dedication to reducing its carbon impact. Intelligent transport technologies entangled knots of inefficiency that had long kept the city hostage and delved deep into the veins of congestion to unravel them. Not only were the journeys shorter, but there was also an emissions reduction achieved as a consequence. As delays in transportation decreased, idle moments disappeared, and the jam of moving cars changed into a smooth, well-choreographed dance. Carbon emissions, which were originally catapulted into the stratosphere by the fervor of inefficiency, have now decreased, much like a symphony reaching a beautiful conclusion (Callahan, 2016).

The Ripple Effects

The effects of this progress in smart travel went much beyond people regular activities. Customers had greater access to local enterprises, which led to their growth. Public health outcomes increased as a result of the enhanced air quality. People and the economy as a whole benefited from the decreased congestion since it resulted in time and money savings. The city's identity saw possibly the most change. It was no longer characterized by gridlock and angry commuters. Instead, it was praised as a Centre of innovation—a location where infrastructure and AI worked together to make cities smarter and more habitable. Customers now have unparalleled access to local firms and enterprises, which was one of this technology advance's dramatic effects. Travelers were effortlessly connected to the social fabric of the place they were visiting thanks to the incorporation of smart travel technologies. This improved accessibility sparked the development and growth of neighborhood businesses, fostering the area's economic vitality. Customers were able to easily discover and interact with local offers, which generated a positive feedback loop between demand and supply and promoted economic growth and vitality. The improvement in air quality came about as a result of a spectacular convergence of technology and public welfare. Smart travel programs reduced traffic congestion and the associated vehicle emissions by using data-driven decision-making and efficient routing. The consequences for public health were directly and favorably impacted by this observable decrease in pollution. As the air grew cleaner, respiratory conditions and other health problems linked to poor air quality also decreased. The overall state of the population improved substantially, with some of this improvement being attributed to the development of smart travel technology (Berglund et al., 2020). Beyond the short-term advantages of lessened traffic and simplified travel, the ramifications of this technological revolution had a significant impact on the economy as a whole. Congestion was reduced by improved traffic flow and clever route design, which resulted in significant time and money savings. Commuters were no longer caught in aggravating traffic jams, which resulted in time savings that could be used for more productive activities. As supply networks improved in efficiency, economic productivity increased concurrently, boosting the region's overall economic performance (Ogie et al., 2017).

Smart Infrastructure and AI: Smart Energy Systems

Building smart energy systems is one area where AI is having a significant influence. Operators can facilitate the switch to renewable energy sources, save customer prices, and greatly enhance efficiency by integrating AI into their energy infrastructure.

Smart energy systems leveraging the latest AI/ML technologies are crucial for modernizing aging infrastructure and supporting the energy needs of 21st century cities in a sustainable manner (Ogie et al., 2017). The field of smart energy systems is one that exemplifies the revolutionary potential of AI in the landscape of contemporary innovation. As operators use AI as a catalyst to usher in an era of renewable energy, cost savings, and exceptional efficiency, the convergence of technology and sustainability weaves a story of transformation. In addition to influencing the present, this symbiotic relationship between AI and energy infrastructure paves the way for a future in which ageing systems are rejuvenated and cities hum with sustainable vitality (Nair et al., 2022). However, the function of AI transcends environmental protection and transforms into an economic boon. A wave of cost reductions spreads across the system as AI-driven insights match energy supply with demand. Customer prices decline, lifting the burden of high bills and making room for cost-effective energy options. It's a transition where technology fills the gap between ecological ideals and financial realities, paving the way for environmentally friendly decisions that are also wise financial decisions (Rasheed, Saad et al., 2021). Efficiency is the ghost at the center of this transformation. Ageing energy infrastructure that was formerly hampered by inefficiency is given new life by AI. Every conduit and switch are given new life by the algorithms, which optimize their operations more precisely than is humanly possible (Ogie et al., 2017).

AI for Demand Prediction and Management

The control and forecast of demand is one of the main uses of AI in the energy sector. Energy providers can more effectively balance supply and demand when short-term load forecasting is accurate. To forecast hourly or sub-hourly demand, AI approaches like repeated artificial neural networks may analyze historical energy consumption data together with elements like weather, time of day, and calendar events (Xiong et al., 2012). These cutting-edge AI networks are essential tools for carefully examining previous energy use trends. These networks achieve a multidimensional analysis that serves as the foundation for precise demand estimates by combining these historical benchmarks with dynamic factors including weather conditions, temporal distinctions, and calendrical events (H. Kim et al., 2021). This makes it possible to purchase power in the wholesale marketplace and schedule-generating resources more effectively. Through dynamic pricing plans, AI is also being used to encourage customers to shift their consumption to off-peak hours, helping to flatten demand curves. AI optimizers can assist to maximize the new demand response potential presented by battery storage and electric cars (O'Dwyer et al., 2019). The use of AI as a dynamic catalyst in this context has numerous distinct aspects. Its seamless integration with wholesale electricity procurement necessitates a methodology that is painstakingly calibrated and driven by real-time insights gained from predictive analytics. The AI-driven algorithms analyze huge pools of historical and current data, supported by sophisticated neural networks, and identify nuanced patterns that elude conventional approaches. With this skill, power suppliers may determine the best times to buy wholesale electricity, assuring a resource-efficient strategy and cost-effective maneuvering in the volatile energy market. In this context, AI acts as a skilled conductor directing the symphony of electric cars and battery storage, giving them a fresh potential for responsive demand management. Electric car charging and energy storage system recedes and flows are identified by AI-powered optimizers, who then dynamically synchronize these systems with changes in demand. With the help of AI-driven algorithms, this synergy creates a complex dance between demand response and resource availability that is guided to work to its fullest capacity (Xiong et al., 2012).

Optimizing the Integration of Renewable Energy

Grid operators have difficulties in maintaining supply-demand equilibrium due to the short-term nature of renewable energy sources like solar and wind. The optimal combination of traditional and renewable energy sources depends heavily on AI. The ideal mix of resources, including those from solar, wind, geothermal, hydro, natural gas, etc., is determined by tools using predictive analytics and optimization algorithms to fulfill load reliably and affordably for each time period (H. Kim et al., 2021). The intelligent tools made possible by predictive analytics and optimization algorithms are at the heart of this dynamic orchestration, revealing the potential for dependability and affordability over a wide range of temporal horizons. Through the lens of predictive analytics, AI gives life to the predictive abilities of these technologies, giving them the capacity to recognize the erratic patterns that characterize the production of renewable energy. Grid operators now have the remarkable ability to predict energy surpluses and shortfalls due to AI algorithms' comprehension of the minute variations in sunshine and wind currents. In conclusion, AI's impact on the energy industry is comparable to that of a maestro conducting a complex symphony and is best exemplified by its role in balancing renewable and conventional energy sources. The baton that directs the ideal symphony of solar, wind, hydro, geothermal, and conventional resources is made up of predictive analytics and optimization algorithms. AI empowers grid operators to choreograph a story of balance, resilience, and affordability that gracefully moves across the temporal tapestry of energy supply and demand through this harmonic blending (H. Kim et al., 2021). According to probable wind and solar projections, AI schedulers may turn on dispatchable generators in advance to make sure there is enough backup power available when renewable output varies. This makes it easier for power systems to integrate more renewable energy into the grid (O'Dwyer et al., 2019). The rise of AI schedulers assumes a crucial position in the tapestry of likely wind and solar forecasts, comparable to expert conductors orchestrating a complicated symphony. These AI-driven virtuosos take front stage when the production of renewable energy oscillates, proactively calling on dispatchable generators to work. This introduction acts as a tactical countermeasure, ensuring that there is a supply of backup power ready to nimbly handle changes in renewable energy (O'Dwyer et al., 2019).

Preventive Maintenance and Anomaly Detection

To forecast equipment breakdowns and suggest preventative maintenance, utilities are implementing AI/ML systems to analyze sensor data from substations, transformers, and other crucial grid assets (Hoffmann et al., 2020). These technologies can assist in preventing expensive unexpected outages by detecting minute changes or irregularities in operation. A grid has a lot of moving parts; thus, AI helps utilities maintain dependability proactively and affordably. Applications include condition-based monitoring of components including feeders, distribution lines, transformers, and circuit breakers. Deep learning algorithms can estimate remaining usable life and maintenance schedules after being educated on past maintenance information (Çınar et al., 2020). In this complex symphony, AI/ML acts as a conductor, guiding the data and insight into a pleasing dance. Each sensor data point is a note in a symphony (Rasheed, Shahzad et al., 2021) of data that AI/ML expertly understands. The algorithms analyze minute differences, undetectable alterations, and aberrations that the human eye could miss, enabling judgment that goes beyond the capabilities of human vision. The inconsistencies or slight variations in operation in the world of grid assets are not considered noise; rather, they are the symphonic notes AI/ML deciphers to foretell equipment breakdowns. In a field where the difference between cost-effective outages and operational stability is razor-thin, AI/ML appears as a powerful defense against the unforeseen. These technologies eliminate the risk of premature breakdowns by anticipatorily spotting irregularities and setting free from the conflict caused by unplanned downtime, where utility functions continue in harmony. The grid represents complexity with its complex component interactions. Similar to master conductors, AI/ML systems analyze this intricacy with the amazing capacity to spot even the smallest dissonance. These technologies analyze the warning indicators, whether they come from a malfunctioning circuit breaker or a misaligned transformer as shown in Figure 9.1, allowing utilities to take action before a catastrophic failure reaches its peak (Krenek et al., 2016).

Figure 9.1: Intelligent infrastructure and systems for smart energy.

Smart Infrastructure and AI: Smart Water Systems

The potential for smart water systems, which combine AI and smart infrastructure, to meet the issues of modern water management is considerable. These systems provide effective resource utilization, early danger identification, and enhanced water quality by utilizing real-time data, predictive analytics, and proactive monitoring. AI technologies provide creative solutions to the problems caused by pollution, ineffective water management practices, and a lack of water (Rojek & Studzinski, 2019). The enormous potential embodied in the symbiotic relationship between AI and intelligent infrastructure, which heralds the arrival of smart water systems, illuminates the horizon of contemporary water management. As represented in Table 9.2, these technologies are prepared to not only address but also overcome the many difficulties that current water management presents. These systems emerge as transformational envoys by cleverly fusing the cognitive power of AI with the complicated web of smart infrastructure. These smart water systems are fundamentally a symphony of efficiency, a skillful balancing act of resource optimization and AI-guided innovation. They use the real-time orchestration of data streams to create a picture of wise resource allocation (Rojek & Studzinski, 2019). One of the main strengths of these systems is their aptitude for early hazard identification. AI emerges as a watchful sentinel in a time when water shortages are a sharp reality. These systems identify abnormalities that may indicate potential dangers, such as leakage, pollution, or abrupt changes in consumption patterns, by interpreting real-time data streams. This foresight serves as a barrier against possible disaster, a proactive step that reduces harm and strengthens resilience. The improvement

Table 9.2: Empowering smart water systems: The synergy of AI and efficient water management

Applications	Importance in smart water systems with AI
Efficient water resource management	AI enables optimization of water distribution, reducing wastage and ensuring equitable supply to different areas. It assists in demand forecasting, balancing supply and demand efficiently.
Real-time monitoring and early warning systems	AI-powered sensors provide real-time data on water levels, flow rates, and potential leaks. Early warning systems use AI algorithms to predict and mitigate water-related emergencies, such as flooding or pipe bursts.
Predictive analytics for water management	AI analyzes historical and real-time data to predict water usage patterns, potential failures, and maintenance needs. It enhances operational efficiency and reduces downtime.
Water quality monitoring and treatment	AI monitors water quality parameters using sensors and data analytics. It identifies contaminants, suggests treatment methods, and optimizes chemical dosages for efficient water treatment.
Integration with IoT and Big Data	AI integrates with IoT devices and collects vast amounts of data. Big Data analysis helps uncover trends, correlations, and insights, enabling better decision-making for water management.

of water quality through proactive monitoring is the distinguishing attribute. Smart sensors and AI's analytical prowess combine to create a continuous feedback loop that unravels the web of water quality parameters (R.A. Ahmed et al., 2022). The core of these systems, this dynamic balance, guarantees that water purity is maintained, avoiding the threat of contamination and assuring a lifeline of safety for communities (K.-G. Kim, 2019). Sensors, data gathering tools, communication networks, AI algorithms, and data analytics tools are frequently found in smart water systems. Together, these elements collect real-time data, keep an eye on water quality, look for leaks, improve water distribution, and make it possible for data-driven decisions (K.-G. Kim, 2019).

Efficient Water Resource Management

By utilizing AI technology to maximize water utilization, cut down on waste, and guarantee water quality, smart water systems enable effective management of water resources. These systems can track water-use trends, spot leaks, and enable preventive maintenance thanks to the integration of cutting-edge sensors and data analytics. AI-powered smart water systems assist save water resources and reducing operating costs by optimizing water distribution networks (Rojek & Studzinski, 2019). The integration of AI technology into the design of smart water systems is redefining the framework for efficient water resource management. The possibility for maximizing water use while reducing waste and protecting water quality emerges as a beacon of transformational innovation in this paradigm shift. These intelligent water systems become skilled stewards of this essential resource, conducting a symphony of accuracy and conservation. These systems, at their heart, represent a dynamic equilibrium—a union of cognitive AI power and the necessity of resource optimization. This partnership creates a landscape in which every water drop is carefully controlled and directed towards utility and sustainability. AI's capacity to glean insights from massive amounts of data and reveal patterns and correlations that control water is what creates the harmony (K.-G. Kim, 2019). These technologies take on waste, a constant foe in the field of water management. They navigate the complex network of pipelines, reservoirs, and distribution systems like watchful sentinels, propelled by the might of cutting-edge sensors. These technologically advanced sensors reveal leakage, irregularities, and inefficiencies that would otherwise go undetected. This ongoing monitoring leads to prompt remedial action, strengthening the base against waste and resource depletion. Through the lens of these systems, water quality—an essential component of public health—is elevated to a level of guarantee. AI continuously evaluates water quality measurements, assessing fluctuations and anomalies that might indicate pollution or deterioration in conjunction with data analytics. The integrity of water resources is ensured by this proactive monitoring, which is supported by real-time information, creating a fabric of safety and well-being for communities (Rojek & Studzinski, 2019).

Real-time Monitoring and Early Warning Systems

The capacity to offer real-time monitoring and early warning systems is one of the key benefits of smart water systems. To identify abnormalities and possible dangers, AI systems examine data gathered from several sensors, including water flow, pressure, and quality. Proactive actions may be made to minimize water loss, reduce environmental concerns, and guarantee the distribution of clean and safe water to communities by identifying leaks, water pollution, or equipment failures early on (Rojek & Studzinski, 2019). The provision of real-time monitoring and the deployment of early warning systems are revolutionary capabilities built into the structure of smart water systems. This crucial aspect demonstrates the connection between cutting-edge technology and the essence of protecting essential water supplies by serving as a vanguard against emerging issues. The profound merging of AI's cognitive abilities with a variety of sensors that map the dimensions of water systems is at the heart of this discovery. These sensors serve as the systems' eyes and ears by being used to record the subtleties of water flow, pressure, and quality. They laboriously collect data streams, each datum containing information that AI carefully interprets. A cacophony of data is transformed into a melodious symphony of insights by this synergy (Rojek & Studzinski, 2019).

Predictive Analytics for Water Management

Predictive analytics powered by AI are essential in smart water systems. AI algorithms can produce precise forecasts of water consumption, resource availability, and future problems by examining historical data, weather patterns, and other pertinent aspects. Informed choices on water allocation, infrastructure improvements, and preventative maintenance may be made by water authorities and utility companies, which improves efficiency and lowers costs (Wang et al., 2006). The alchemy of AI algorithms, designed to glean significant insights from a nuanced combination of historical data, weather patterns, and contextual nuances, is at the core of this shift. These algorithms explore the data landscape much like an expert mapper might, mapping a path beyond the fleeting present and into the world of prediction. AI uncovers patterns that are invisible to conventional research by closely examining previous behavior and water use trends. These algorithms mimic the complex choreography of weather patterns by including meteorological variables and tying their effect to water dynamics. The combination increases prediction accuracy by adding aspects like wind, temperature, and precipitation to the story (K.-G. Kim, 2019).

Water Quality Monitoring and Treatment

For the sake of public health, maintaining water quality is essential. AI algorithms are used by smart water systems to continually track water quality indicators including pH, turbidity, and chemical composition. Real-time data analysis makes it possible to identify toxins or unusual fluctuations early, setting off the proper water treatment procedures and guaranteeing regulatory compliance (Wang et al., 2006). A fundamental cornerstone of public health that transcends time and location is ensuring the cleanliness of water. The introduction of AI algorithms inside smart water systems reveals a transformational paradigm in this critical endeavor, one that has the ability to protect communities from the specter of pollution. Water quality becomes more than just an asset by combining technology and attention, becoming a sentinel of well-being (Oad et al., 2023). The power of AI algorithms, which are similar to perceptive sentinels that navigate the fluid world of water systems, is at the heart of this shift. Their task is to monitor and closely examine the chemical composition, turbidity, and pH indicators that speak of water quality (K.-G. Kim, 2019).

Integration with IoT and Big Data

The IoT and big data technologies are used by smart water systems to further improve their capabilities. These systems may optimize water distribution, simplify procedures, and foster customer participation by connecting with IoT devices like smart meters and remote-controlled valves. Big data analytics allow for the extraction of useful insights from enormous volumes of data, allowing long-term planning and evidence-based decision-making (Huseien & Shah, 2022). The seamless integration of IoT devices into smart water systems is at the core of these developments. These systems can now achieve previously unheard-of levels of engagement, agility, and optimization because of this connection. These systems become dynamic ecosystems that react in real-time to changes in water demand and supply by integrating with IoT devices like smart meters and remote-controlled valves (H. Kim et al., 2021). The distribution of water is optimized, which is one of the main transforming elements. Smart water systems are able to monitor water flow, pressure, and consumption patterns with outstanding precision through real-time data capture and analysis. The systems can dynamically change distribution networks thanks to this dynamic monitoring, which successfully reduces leaks, reduces waste, and ensures that water resources are used as efficiently as possible (Şerban & Lytras, 2020).

Smart Infrastructure and AI: Smart Buildings

By utilizing AI-driven systems to optimize energy usage in line with global environmental goals, smart buildings play a crucial role in tackling energy efficiency and sustainability issues. This technological integration fosters increased productivity by supporting the creation of pleasant settings that encourage well-being while also saving a significant amount of money through resource allocation and operational efficiency. Additionally, AI-driven security systems improve safety by identifying and thwarting threats, and obtaining green building certifications increases a building's worth and draws in eco-aware tenants. By acting as a base for new technologies and being responsive to changing demands and developments, these smart buildings also assure future preparedness (Mehmood et al., 2020). Intelligent AI-driven technologies that support smart buildings act as catalysts in the worldwide quest for sustainability and energy efficiency. These engineering wonders aren't just buildings; they're interconnected ecosystems that balance environmental stewardship, operational excellence, and comfort for people. Smart buildings cross the nexus of productivity, well-being, cost-efficiency, security, and future-readiness in this union of technology and purpose. The coordination of energy optimization in line with global environmental ambitions is at the core of this transition. Here, AI excels, orchestrating the perfect equilibrium between energy consumption and conservation. These systems maintain an ongoing conversation with energy networks and adjust in real-time to changes in supply and demand, reducing (Şerban & Lytras, 2020).

Internet of Things (IoT) Integration

Smart buildings integrate a complex network of sensors and gadgets that are placed strategically all throughout the building to capture the power of the IoT. These sensors carefully gather a wide variety of data, such as but not limited to occupancy patterns, ambient temperature, lighting levels, air quality measurements, and energy consumption trends. This network of interconnected devices creates a steady stream of data that serves as the foundation for the building's intelligent operations. Similar to the building's senses, these sensors create a watchful network that reaches every crevice. They perform more than just data collecting; they also serve as the building's windows into the outside world. Precision is used to record occupancy patterns, which are similar to the ups and downs of a human heartbeat. A comfortable and productive workplace is created by adjusting the ambient temperature and lighting settings. Measurements of air quality act as well-being indicators, making sure that people are breathing healthy air. Similar to how a pulse is tracked, energy consumption trends are studied to maximize resource use and reduce waste (Huseien & Shah, 2022).

Data Analytics and AI

To extract useful insights, the obtained data is analyzed using AI and sophisticated analytics techniques. To maximize energy efficiency and comfort levels, AI systems may automatically change temperature, lighting, and other settings based on user preferences, weather, and occupancy patterns (Luckey et al., 2021). Data is transformed into a tapestry of insights that are woven into the complex fabric of smart buildings by the delicate hands of AI and advanced analytics. Here is where information's genuinely transforming power manifests itself. As it moves through the analysis rooms, the data gathered by the sensor network has a life of its own and produces a symphony of knowledge that directs the building's operations. Advanced analytics methods cast a critical eye on the data streams as they enter the repository of AI-driven systems. These methods reveal the layers of intricacy inside the data, just like professional interpreters do. They raise the status of raw data to a level of valuable insights by revealing patterns, correlations, and trends that may be obscure to the unaided eye (Luckey et al., 2021).

Energy Management

To monitor and regulate energy use, smart buildings use intelligent energy management systems. IoT sensor data is analyzed by AI algorithms to pinpoint inefficiencies and suggest energy-saving strategies. In addition to lowering carbon impact, this also results in significant financial savings. Through the fusion of intelligence and innovation, the idea of managing energy goes through a significant transition in the age of smart buildings (W.S. Alaloul et al., 2020). The implementation of intelligent energy management systems, which serve as the architects of effectiveness, sustainability, and financial restraint, is at the center of this progress. The synchronized synergy between IoT sensors and AI algorithms—a relationship that traverses the spheres of data, insight, and action—accomplishes this (O'Dwyer et al., 2019).

Security and Safety

AI-driven security systems with video analytics, facial recognition, and behavior tracking offer improved surveillance. Smart buildings can recognize possible risks, spot odd activity, and issue alerts in real-time, strengthening overall safety measures (Berglund et al., 2020).

Occupant Comfort and Experience

By using customized settings and automation, smart buildings aim to enhance comfort for passengers. Individual preferences are learned by AI algorithms, which then modify ambient parameters like lighting, temperature, and others in accordance. Additionally, users may communicate with the building's systems through voice-activated assistants or smartphone applications. A mutually beneficial conversation between sensors, data, and AI-driven insights is the end outcome. This discussion focuses on identifying energy inefficiencies, places where electricity is wasted or improperly distributed. With this well-informed perspective, AI algorithms take on the role of strategic advisers and propose unconventional energy-saving tactics. Optimizing lighting schedules, coordinating temperature management with occupancy patterns, or carefully allocating electricity during peak hours are a few examples of these tactics. It is an intelligent choreography that redefines energy management. This choreographed dance has several ramifications. This symphony results in measurable decreases in carbon footprint, above and beyond the spheres of effectiveness and prudent financial management. Through the lens of smart building intelligence, energy conservation is a commitment to the larger story of sustainability—a story that connects with the ambitions of the entire world for the environment (Saad, Salah Alaloul et al., 2020).

References

Adochiei, F.-C., Nicolescu, Ş.-T., Adochiei, I.-R., Seritan, G.-C., Enache, B.-A., Argatu, F.-C. and Costin, D. (2020). Electronic System for Real-time Indoor Air Quality Monitoring. *2020 International Conference on E-Health and Bioengineering (EHB)*, 1–4. https://doi.org/10.1109/EHB50910.2020.9280192.

Ahmad, T., Madonski, R., Zhang, D., Huang, C. and Mujeeb, A. (2022). Data-driven probabilistic machine learning in sustainable smart energy/smart energy systems: Key developments, challenges, and future research opportunities in the context of smart grid paradigm. *Renewable and Sustainable Energy Reviews*, 160, 112128. https://doi.org/https://doi.org/10.1016/j.rser.2022.112128.

Ahmed, R.A., Hemdan, E.E.-D., El-Shafai, W., Ahmed, Z.A., El-Rabaie, E.-S.M. and Abd El-Samie, F.E. (2022). Climate-smart agriculture using intelligent techniques, blockchain, and Internet of Things: Concepts, challenges, and opportunities. *Transactions on Emerging Telecommunications Technologies*, 33(11), e4607. https://doi.org/https://doi.org/10.1002/ett.4607.

Ahmed, S., Khitab, A., Mehmood, K. and Tayyab, S. (2020). Green non-load bearing concrete blocks incorporating industrial wastes. *SN Applied Sciences*, 2(2). https://doi.org/10.1007/s42452-020-2043-6 WE - Emerging Sources Citation Index (ESCI).

Alaloul, W.S., Saad, S. and Qureshi, A.H. (2020). Construction sector: IR 4.0 applications. *In:* C.M. Hussain and P. Di Sia (Eds.), *Handbook of Smart Materials, Technologies, and Devices: Applications of Industry 4.0*, pp. 1–50. Springer International Publishing. https://doi.org/10.1007/978-3-030-58675-1_36-1.

Ali, A., Ud-Din, S., Saad, S., Ammad, S., Rasheed, K. and Ahmad, F. (2021). Artificial Neural Network Approach to Study the Effect of Driver Characteristics on Road Traffic Accidents. *2021 International Conference on Data Analytics for Business and Industry, ICDABI 2021*. https://doi.org/10.1109/ICDABI53623.2021.9655827.

Badshah, I., Khan, Z., Saad, S., Khan, F., Aslam, S. and Khattak, K. (2020). A videogrammetric analysis of on peak/off peak traffic density: A case of Board Bazaar Peshawar. *Pakistan Journal of Engineering and Technology*, 3(03 SE-Articles). https://doi.org/10.51846/vol3iss03pp38-45.

Berglund, E.Z., Monroe, J.G., Ahmed, I., Noghabaei, M., Do, J., Pesantez, J.E., Khaksar Fasaee, M.A., Bardaka, E., Han, K. and Proestos, G.T. (2020). Smart infrastructure: A vision for the role of the civil engineering profession in smart cities. *Journal of Infrastructure Systems*, 26(2), 3120001.

Bhogaraju, S.D. and Korupalli, V.R.K. (2020). Design of smart roads: A vision on Indian smart infrastructure development. *2020 International Conference on Communication Systems & Networks (COMSNETS)*, 773–778.

Callahan, D. (2016). The five horsemen of the modern world. *In: Climate, Food, Water, Disease, and Obesity*. Columbia University Press. https://doi.org/doi:10.7312/call17702.

Çınar, Z.M., Abdussalam Nuhu, A., Zeeshan, Q., Korhan, O., Asmael, M. and Safaei, B. (2020). Machine learning in predictive maintenance towards sustainable smart manufacturing in Industry 4.0. *Sustainability*, 12(19). https://doi.org/10.3390/su12198211.

Curley, M. and Salmelin, B. (2018). Sustainable intelligent living. *In:* M. Curley and B. Salmelin (Eds.), *Open Innovation 2.0: The New Mode of Digital Innovation for Prosperity and Sustainability*, pp. 27–38. Springer International Publishing. https://doi.org/10.1007/978-3-319-62878-3_3.

Hoffmann, M.W., Wildermuth, S., Gitzel, R., Boyaci, A., Gebhardt, J., Kaul, H., Amihai, I., Forg, B., Suriyah, M., Leibfried, T., Stich, V., Hicking, J., Bremer, M., Kaminski, L., Beverungen, D., zur Heiden, P. and Tornede, T. (2020). Integration of novel sensors and machine learning for predictive maintenance in medium voltage switchgear to enable the energy and mobility revolutions. *Sensors*, 20(7). https://doi.org/10.3390/s20072099.

Hoosain, M.S., Paul, B.S. and Ramakrishna, S. (2020). The impact of 4IR digital technologies and circular thinking on the United Nations sustainable development goals. *Sustainability*, 12(23). https://doi.org/10.3390/su122310143.

Huseien, G.F. and Shah, K.W. (2022). A review on 5G technology for smart energy management and smart buildings in Singapore. *Energy and AI*, 7, 100116.

Kim, H., Choi, H., Kang, H., An, J., Yeom, S. and Hong, T. (2021). A systematic review of the smart energy conservation system: From smart homes to sustainable smart cities. *Renewable and Sustainable Energy Reviews*, 140, 110755.

Kim, K.-G. (2019). Development of an integrated smart water grid model as a portfolio of climate smart cities. *Journal of Smart Cities*, 3(1), 23–34.

Krenek, J., Kuca, K., Blazek, P., Krejcar, O. and Jun, D. (2016). Application of artificial neural networks in condition based predictive maintenance. *In:* D. Król, L. Madeyski and N.T. Nguyen (Eds.), *Recent Developments in Intelligent Information and Database Systems*, pp. 75–86. Springer International Publishing. https://doi.org/10.1007/978-3-319-31277-4_7.

Liu, Z. and Zhang, L. (2020). A review of failure modes, condition monitoring and fault diagnosis methods for large-scale wind turbine bearings. *Measurement*, 149, 107002. https://doi.org/https://doi.org/10.1016/j.measurement.2019.107002.

Luckey, D., Fritz, H., Legatiuk, D., Dragos, K. and Smarsly, K. (2021). Artificial intelligence techniques for smart city applications. *Proceedings of the 18th International Conference on Computing in Civil and Building Engineering: ICCCBE 2020*, 3–15.

Mehmood, R., Katib, S.S.I. and Chlamtac, I. (2020). *Smart Infrastructure and Applications*. Springer.

Menouar, H., Guvenc, I., Akkaya, K., Uluagac, A.S., Kadri, A. and Tuncer, A. (2017). UAV-enabled intelligent transportation systems for the smart city: Applications and challenges. *IEEE Communications Magazine*, 55(3), 22–28.

Nair, A.K., John, C. and Sahoo, J. (2022). Implementation of Intelligent IoT. *In:* Z. Boulouard, M. Ouaissa, M. Ouaissa and S. El Himer (Eds.), *AI and IoT for Sustainable Development in Emerging Countries: Challenges and Opportunities*, pp. 27–50. Springer International Publishing. https://doi.org/10.1007/978-3-030-90618-4_2.

Nasr, M.S., Ali, I.M., Hussein, A.M., Shubbar, A.A., Kareem, Q.T. and Abdul Ameer, A.T. (2020). Utilization of locally produced waste in the production of sustainable mortar. *Case Studies in Construction Materials*, 13. https://doi.org/10.1016/j.cscm.2020.e00464 WE - Science Citation Index Expanded (SCI-EXPANDED).

O'Dwyer, E., Pan, I., Acha, S. and Shah, N. (2019). Smart energy systems for sustainable smart cities: Current developments, trends and future directions. *Applied Energy*, 237, 581–597.

Oad, V.K., Szymkiewicz, A., Khan, N.A., Ashraf, S., Nawaz, R., Elnashar, A., Saad, S. and Qureshi, A.H. (2023). Time series analysis and impact assessment of the temperature changes on the vegetation and the water availability: A case study of Bakun-Murum Catchment Region in Malaysia. *Remote Sensing Applications: Society and Environment*, 29, 100915. https://doi.org/https://doi.org/10.1016/j.rsase.2022.100915.

Ogie, R.I., Perez, P. and Dignum, V. (2017). Smart infrastructure: An emerging frontier for multidisciplinary research. *Proceedings of the Institution of Civil Engineers – Smart Infrastructure and Construction*, 170(1), 8–16.

Plageras, A.P., Psannis, K.E., Stergiou, C., Wang, H. and Gupta, B.B. (2018). Efficient IoT-based sensor BIG Data collection – Processing and analysis in smart buildings. *Future Generation Computer Systems*, 82, 349–357. https://doi.org/https://doi.org/10.1016/j.future.2017.09.082.

Qureshi, A.H., Alaloul, W.S., Manzoor, B., Musarat, M.A., Saad, S. and Ammad, S. (2020). Implications of Machine Learning Integrated Technologies for Construction Progress Detection under Industry 4.0 (IR 4.0). *2020 Second International Sustainability and Resilience Conference: Technology and Innovation in Building Designs (51154)*, 1–6. https://doi.org/10.1109/IEEECONF51154.2020.9319974.

Rasheed, K., Saad, S., Shahzad, L., Ammad, S., Ali, A. and Badshah, I. (2021). Application of Gesture Data Recognition in a Human-Interactive Leap Motion Sensor Chair. *2021 International Conference on Data Analytics for Business and Industry, ICDABI 2021*. https://doi.org/10.1109/ICDABI53623.2021.9655819.

Rasheed, K., Shahzad, L., Saad, S., Khan, H.A., Ahmed, W. and Sadiq, T. (2021). Parking Guidance System Using Wireless Sensor Networks. *2021 International Conference on Decision Aid Sciences and Application, DASA 2021*. https://doi.org/10.1109/DASA53625.2021.9682213.

Rojek, I. and Studzinski, J. (2019). Detection and localization of water leaks in water nets supported by an ICT system with artificial intelligence methods as a way forward for smart cities. *Sustainability*, 11(2), 518.

Saad, S., Alaloul, W.S., Ammad, S., Qureshi, A.H., Altaf, M. and Rasheed, K. (2020). Design Phase Carbon Emission Prediction Using a Visual Programming Technique. *2020 2nd International Sustainability and Resilience Conference: Technology and Innovation in Building Designs*. https://doi.org/10.1109/IEEECONF51154.2020.9319978

Saad, S., Salah Alaloul, W., Ammad, S., Hannan Qureshi, A., Mohsen Mohammed Alawag, A., Kumar Oad, V. and Altaf, M. (2020). A noded carbon emission tool (NCET) to measure embodied carbon (EC) in compositional cementitious materials. *Solid State Technology*, 63(6), 4099–4105.

Saad, S., Alaloul, W.S. and Ammad, S. (2022). Role of cyber-physical systems in smart cities. *In:* D.W.S. Alaloul (Ed.), *Cyber-Physical Systems in the Construction Sector* (1st Edn.), p. 19. CRC Press. https://doi.org/9781003190134.

Şerban, A.C. and Lytras, M.D. (2020). Artificial intelligence for smart renewable energy sector in europe—Smart energy infrastructures for next generation smart cities. *IEEE Access*, 8, 77364–77377.

Shahid, N., Shah, M.A., Khan, A., Maple, C. and Jeon, G. (2021). Towards greener smart cities and road traffic forecasting using air pollution data. *Sustainable Cities and Society*, 72, 103062. https://doi.org/https://doi.org/10.1016/j.scs.2021.103062.

Wang, F.-Y., Zeng, D. and Yang, L. (2006). Smart cars on smart roads: An IEEE intelligent transportation systems society update. *IEEE Pervasive Computing*, 5(4), 68–69.

Xiong, Z., Sheng, H., Rong, W. and Cooper, D.E. (2012). Intelligent transportation systems for smart cities: A progress review. *Science China Information Sciences*, 55, 2908–2914.

Yigitcanlar, T., Mehmood, R. and Corchado, J.M. (2021). Green artificial intelligence: Towards an efficient, sustainable and equitable technology for smart cities and futures. *Sustainability*, 13(16). https://doi.org/10.3390/su13168952.

CHAPTER

10

Ethical and Societal Implications of AI in Construction

Kumeel Rasheed, Muhammad Waqas, Syed Saad, Ahmad Zaland and Zawar Ali

Ethical Considerations of AI-assisted Construction

The construction industry faces challenges, including a shortage of workers, safety concerns, limited productivity, and the need, for sustainability. To address these challenges embracing technologies such as artificial intelligence (AI) robotics, Building Information Modeling (BIM), and the Internet of Things (IoT) is crucial (Abioye et al., 2021). However, implementing these innovations requires a sense of responsibility. Since the construction industry provides employment opportunities' changes in jobs due to technology, it requires consideration to avoid negative impacts on society (George et al., 2012). As these technologies streamline operations, certain traditional roles may transform. It is estimated that within the 15 years around 30% of onsite tasks like material handling and assembly could be automated (Pan & Zhang, 2021). While new support positions may emerge alongside automated systems, existing job roles are at risk unless accompanied by strategies. Companies introducing these technologies have an obligation to conduct assessments of their impact and explore models that combine human skills with machine capabilities before making large-scale workforce changes (Issa et al., 2016). Engaging with labor unions is vital in creating negotiated plans that include retraining opportunities, new job prospects and job security, over agreed timeframes (Osterman et al., 2002).

As an illustration, the Danish business MT Haggard started a robotics initiative by consulting universities to ascertain how roles were altering (Yermack, 2010). After that, they invested two years in educating welders, electricians, and carpenters to manage and operate robots. In this changing environment, governments and educational institutions also have important duties. Together, well designed lifelong learning programs, strong worker rights, retraining income support mechanisms, and active labor programs help to secure inclusive transitions, with an emphasis on protecting the workforce's most vulnerable groups. The government of Singapore's CPCB program offers subsidies and allowances as incentives to encourage small and medium-sized businesses to adopt new technologies. It also enables worker reskilling through a variety of vocational courses that cover topics like computer-aided design (CAD), drones, and robotics (Moniz & Krings, 2016). Technology developers, who are the main drivers of innovation, must assume early analyses of labor consequences and form alliances with end users to gradually deploy these tools while utilizing already-existing skill sets. To reduce the possibility of joblessness, European BIM vendor Prolitas conducted social assessments of computer vision quality inspection modules in partnership with a safety organization prior to market debut. Assistant roles were initially given priority rather than complete replacements. An increasing number of people are becoming concerned about algorithmic transparency as AI systems begin to perform operational and strategic planning tasks. As a lack of transparency in 'black-box' systems can result in unintended biases and effects, stakeholders rightfully want to understand how AI influences important decisions. To foster trust, vendors must follow the "ethical by design" principles and do thorough impact analyses. This strategy is best demonstrated by the Chinese BIM company Zhaomei, which incorporates explainable approaches in structural fault detection software and gives users confidence scores and step-by-step justification for alarms (Jiang et al., 2015). Although technology providers create these game-changing capabilities, construction companies are ultimately responsible for their implementation. Establishing vendor evaluation standards, extensive use-case documentation, bias identification processes, recourse mechanisms, ongoing audits, and compulsory reporting are all necessary for effective governance. In Europe, Vinci is used as an example because it established an AI committee after a predictive maintenance pilot and gave it the crucial responsibility of reviewing provider protocols, supervising new deployments, scrutinizing anomalies, monitoring workforce interactions, and advocating for end-user needs. Governments are crucial in developing policy frameworks as AI grows on a national scale. Utilizing the expertise of experts, multi-stakeholder processes help to create consensus-based recommendations that cover topics like sector-specific guidance, regulatory transparency benchmarks, coordinated oversight mechanisms, and standardized contractual clauses that codify best practices. The initiative to create a national AI accountability framework, ensuring the responsible application of these technologies throughout project lifecycles, is exemplified by Singapore's establishment of a Construction Governance Council in 2021, which is composed of representatives from regulatory bodies, educational institutions, major firms, and labor unions (Xia et al., 2023). In conclusion, cutting-edge technology unquestionably ushers in fundamental changes in the landscape of talents within the building industry. To protect inclusive transitions and put protections in place for increasingly autonomous systems that have a significant impact on livelihoods and infrastructure, proactive engagement between corporations, governments, technology developers, and worker groups becomes essential. Digital tools have the potential to increase productivity while respecting the socioeconomic interests of the employees, business owners, and communities that the construction sector devotedly serves (Ammad, Saad et al., 2021). This is possible with concentrated efforts and a shared vision.

Ethics of Job Disruption and Skills Transition

Approximately 25% of work tasks in the construction industry could be automated, according to a McKinsey Global Institute paper on the subject. While AI-powered devices like computer vision, robots, and drones have the ability to perform risky or monotonous activities, they also raise worries about job displacement, the need for retraining, and the importance of transition support for workers. Before implementing new technology, construction companies must do rigorous analyses to see how they will impact current positions and handle these ethical quandaries. To properly collaborate with AI systems in a hybrid human-machine teaming model, workers may need to be reskilled rather than having their jobs completely replaced by machines (Qureshi et al., 2020). Governments are essential in creating policies that promote skill development and facilitating reskilling and upskilling programs, frequently in collaboration with educational institutions or receiving financial support for vocational training. Additionally, specific aid should be given to those who might lose their jobs as a result of technological improvements. To ensure a fair transition for all impacted employees when jobs change, technology developers must jointly examine the risk for employment disruption with businesses and labor unions. Through negotiated transfer procedures, redeployment opportunities, and access to lifelong learning, construction unions can serve as crucial allies in promoting worker interests. For instance, the Scandinavian construction company Skanska worked closely with the unions to smoothly transition welders and assemblers into new quality control roles supervising automated processes for the deployment of 3D printed mini-homes in the UK. In addition, Singapore has taken initiative in this area, initiating the "Construction Singapore Development" program in 2021 and providing $130 million to stimulate the adoption of robotics and automation in the construction industry. For employees who reskill in cutting-edge industries like BIM, precast concrete, and digital fabrication, this program offers on-the-job training allowances (Saad et al., 2022). In essence, the construction sector is on the verge of a huge technological shift, making it critical to address ethical issues relating to worker retraining, displacement, and transition support. To negotiate this change and guarantee a just and successful future for all stakeholders in the construction sector, coordination between construction companies, governments, technology innovators, and labor unions is encouraged (Aldoseri et al., 2023).

Societal Implications of AI in Construction

The built environment and building community ties are two important goals of the construction business. It is crucial to fully understand the societal implications of this technological growth as AI, robotics, drones, and digital twins usher in dramatic changes within the sector (Mourtzis et al., 2022). While these technologies successfully handle a variety of business difficulties, their significant impacts on accessibility, labor, communities, and sustainable growth call for proactive attention to guarantee that the advantages for various stakeholders are in line with the UN's development goals. It's crucial to recognize that dangers are a part of any construction job in the globe. While the safe application of AI and robotics will surely improve safety, changing job requirements may disproportionately burden some workforce segments. Governments, businesses, and unions all have a part to play in this situation. Governments protect workers by enacting laws like minimum wages and actively promoting reskilling programs, and unions negotiate just transition agreements that cover income protection, upskilling, and the creation of new opportunities. The effects on disadvantaged populations must also be carefully considered by technology developers, who must also work together to provide inclusive solutions. Exemplary cases highlight progressive strategies, such as Singapore's initiatives that support SMEs (small and medium enterprises) in adopting new technology while championing hiring equity, and European businesses working with unions to jointly design reskilling programs via modular mobile courses, strengthening safety certifications and digital skills. It is crucial to remove any biases in datasets and guarantee the active participation of stakeholders as these cutting-edge technologies shape our built environments and have an impact on people' lives. Failure to do so may result in the absence of socio-environmental aspects from AI-driven urban planning, which may marginalize accessibility and the aspirations of inhabitants who are less advantaged. To handle these disruptive shifts carefully and ethically, transparent governance frameworks and constructive engagement across all sectors are essential.

Job Quality, Reskilling, and an Inclusive Labor Market

Although they are a common source of employment worldwide, construction occupations frequently involve risks. The proper application of robotics and AI in this sector has the potential to improve safety while also changing the skill sets needed. The burdens of this transformation are not equally dispersed, and the emergence of "AI jobs" does not imply that everyone involved will have stability or respectable working circumstances. By establishing measures like minimum salaries, comprehensive benefits, encouraging collective worker representation, and adopting proactive labor regulations, governments play a crucial role in protecting against excessive precarity. Equally crucial is promoting progressive contracting methods that reward the adoption of high-quality jobs. To establish fair transition agreements that cover worker reskilling and upskilling projects, transferrable credentials, income protections, and access to the new opportunities brought about by these disruptive technologies, businesses and labor unions should cooperate. To co-create solutions that prioritize inclusion, technology developers must be accountable for determining the potential effects on vulnerable groups, such as migrant workers, and actively interact with affected worker communities. For instance, Singapore's Construction Productivity and Capacity Building program offers small- and medium-sized contractors essential help by subsidizing the adoption of new technology, while its Labour Foundation encourages ethical recruiting procedures and uniform training across the sector. Construction 3D, a European software company, gives an example by collaborating with unions that represent masons and carpenters to create augmented reality job aids that improve trainees' marketable digital skills and safety certifications via modular mobile courses. Governments also play a part in ensuring that policies encourage the development of skills and support reskilling and upskilling initiatives through educational institutions or vocational training subsidies, with a focus on specialized aid for displaced employees (Panth & Maclean, 2020). Overall, as roles continue to change, these transformational technologies have an ethical responsibility to examine their potential to disrupt employment and work with businesses and unions to ensure a fair transition for all impacted workers. Construction unions can effectively represent worker interests by taking part in negotiated transition procedures, providing possibilities for redeployment, and facilitating access to lifelong learning opportunities in collaboration with businesses and governments (Mitchell, 1998).

Community Engagement, Accessibility, and Sustainable Urban Development

The use of AI and digital tools in infrastructure design, construction, and urban planning has a significant influence on entire communities, influencing their built environments. While biases in training data and a lack of meaningful stakeholder participation pose inherent dangers, data-driven techniques have the potential to address sustainability concerns, improve mobility access, boost resilience, and promote livability on a wide scale. It can further disadvantage marginalized groups to disregard the various interests of communities, particularly when it comes to the distribution of public resources like transit systems and green spaces. Predictive urban modelling may also favor short-term economic indicators over long-term socio-environmental consequences if sufficient accountability is lacking. The establishment of AI governance frameworks that require transparent and significant public participation processes during the development and deployment of technological solutions that have a direct influence on communities is a critical obligation of the government. Equally important are safeguards for personal information and control systems to address any unforeseen social, economic, or environmental effects. Construction companies can create community benefits agreements that reflect consensus on the needs of underserved groups, including them directly in data gathering processes whenever possible to assure representativeness, while testing AI tools in community contexts (Mohsen Alawag et al., 2023). Academic research consortiums that work together to evaluate the long-term effects of AI-driven urban planning strategies provide a multidisciplinary viewpoint that strikes a balance between technical, economic, and human issues. Accountability is ensured via participatory action models, which call for ongoing cooperation with civic stakeholders to track results after deployment. As an illustration, Singapore's AI for Sustainable Urban Mobility and Planning program serves as a paradigm, bringing together input from business, the government, and non-profits to simulate constructed environments that balance profitability, accessibility to transportation, and environmental stewardship (Yigitcanlar & Cugurullo, 2020). Similar to this, American developer Bechtel's use of digital modeling for planned transportation projects in Virginia stands out since it was motivated by comprehensive community participation that takes into consideration various quality of life preferences.

Addressing the Challenges and Risk of AI in Construction

The construction industry is not an exception to how AI has transformed many other industries. The use of AI in construction might completely change how projects are organized, managed, and carried out. AI has a wide range of potential advantages for the construction sector, from boosting safety measures to optimizing resource allocation. To guarantee that AI is successfully incorporated into construction, however, these potentials come with considerable risks and obstacles that must be handled. The difficulties and dangers of using AI in building will be examined in this article, along with mitigation techniques.

Challenges in Implementing AI in Construction

Data Management and Quality

Data management and quality are two of the key difficulties in applying AI in construction. The training, learning, and decision-making processes of AI systems are significantly reliant on data (Abbasi et al., 2022). Data from architectural blueprints, project timetables, material specifications, and equipment data are all generated in great quantities throughout construction projects (Ammad, Alaloul et al., 2021). The majority of this data, however, is unorganized, inconsistent, or lacking. It is vital to make certain that the data utilized to train AI models is precise and correct. The efficacy of AI solutions can be weakened by bad data quality, which can provide biased or incorrect predictions (Teng et al., 2021).

Construction businesses must make investments in data management and quality assurance procedures to overcome this issue. To guarantee that AI systems have access to high-quality data, this may entail data cleansing, standardization, and validation. Furthermore, businesses may use tools like data analytics and machine learning to spot and immediately fix data problems.

Integration with Existing Systems

Many construction companies already have established systems and workflows in place. Integrating AI into these existing processes can be challenging, as it often requires compatibility with legacy systems and technologies. Incompatibility issues can result in disruptions to project timelines and increased costs.

To overcome this challenge, construction firms should carefully evaluate their current systems and identify areas where AI can be seamlessly integrated. Customized solutions and APIs (application programming interfaces) can be developed to bridge the gap between AI applications and existing systems. Collaboration with technology providers and consultants specializing in AI integration can also be beneficial (Perifanis & Kitsios, 2023).

Cost of Implementation

The initial cost of implementing AI in construction can be substantial. Procuring AI hardware, software, and training personnel can strain the budgets of smaller construction firms. Moreover, there may be hidden costs associated with system maintenance, upgrades, and ongoing support.

To address this challenge, construction companies should conduct a thorough cost-benefit analysis to assess the long-term advantages of AI adoption. While the upfront costs may be high, AI can deliver substantial returns on investment through improved project efficiency, reduced labor costs, and enhanced decision-making. Companies can also explore partnerships or collaborations to share the financial burden of AI implementation (Kiriiri et al., 2020).

Skilled Workforce Shortage

A professional workforce with knowledge of AI technologies, data science, and ML is needed for the effective adoption of AI in the construction industry. But because there aren't enough people with these specialized abilities, it's hard for construction businesses to find and keep the expertise needed to create and maintain AI systems.

Construction companies might engage in training programs and collaborations with educational institutions to upskill their current personnel to meet this difficulty. They can also use outsourcing and cooperation with AI service providers to gain temporary access to the necessary skills (Fountaine et al., 2019).

Safety and Ethical Concerns

AI applications in construction can raise safety and ethical concerns, particularly in situations where AI systems make critical decisions. For example, autonomous construction equipment may raise questions about operator safety and liability in the event of accidents. There are also ethical concerns related to data privacy and security, as construction projects often involve sensitive information.

To mitigate safety and ethical concerns, construction companies should prioritize safety standards and guidelines for AI-driven equipment. This includes ensuring that autonomous machines have robust safety features and fail-safes to prevent accidents. Additionally, companies should implement data protection measures, including encryption and access controls, to safeguard sensitive project data (Shneiderman, 2020).

Resistance to Change

Resistance to change is a common challenge when introducing AI into any industry, including construction. Employees and stakeholders may be hesitant to adopt new technologies, fearing job displacement or disruptions to established processes. This resistance can hinder the successful implementation of AI solutions.

To address this challenge, construction companies should invest in change management strategies that involve employees in the decision-making process and provide training and support during the transition. Clear communication about the benefits of AI, including how it can enhance job roles and create new opportunities, is essential in overcoming resistance (Dwivedi et al., 2021).

Risks Associated with AI in Construction

While AI offers numerous benefits, there are also inherent risks associated with its use in construction:

Unreliable Predictions: The quality of AI models depends on the data they are trained on. AI forecasts may not be accurate if the data is incorrect, skewed, or unrepresentative. Prediction errors in the construction industry can result in expensive errors, project delays, and safety risks.

Mitigation Strategy: Construction companies should invest in data quality and validation processes to ensure that the data used for AI training is accurate and unbiased. Regular monitoring and validation of AI predictions against real-world outcomes can help identify and rectify reliability issues (Hamouda, 2023).

Cybersecurity Threats

AI applications in construction often involve the collection and storage of sensitive project data, making them attractive targets for cyberattacks. Breaches in cybersecurity can result in data theft, project disruptions, and financial losses.

Mitigation Strategy: Construction companies should prioritize cybersecurity measures, including encryption, firewalls, and intrusion detection systems, to protect AI systems and project data. Regular security audits and employee training on cybersecurity best practices are also essential (de Soto et al., 2020).

Lack of Regulation

The use of AI in construction is a relatively new frontier, and regulatory frameworks may lag behind technological advancements. This can create legal uncertainties and liability issues, especially in cases where AI systems make critical decisions.

Mitigation Strategy: Construction companies should stay informed about evolving AI regulations and work with legal experts to ensure compliance with existing laws. They should also establish clear liability and accountability frameworks for AI-driven processes and decisions (Weber-Lewerenz, 2021).

Overreliance on AI

Overreliance on AI systems without human oversight can lead to complacency and a decreased ability to respond to unexpected situations. Construction projects often encounter unforeseen challenges that require human judgment and adaptability (Madni & Jackson, 2009).

Mitigation Strategy: Construction companies should use AI as a tool to augment human decision-making rather than replace it entirely. Maintaining a balance between AI automation and human oversight is crucial for effective risk management.

Environmental Impact

AI-driven construction processes, particularly those involving autonomous machinery, can have environmental impacts. Energy consumption, resource utilization, and emissions need to be carefully managed to minimize the environmental footprint of AI in construction.

Mitigation Strategy: Construction companies should assess the environmental impact of AI technologies and prioritize sustainability in their AI-driven processes. This may involve using energy-efficient equipment, optimizing resource allocation, and monitoring emissions (Chen et al., 2023).

Strategies to Address Challenges and Mitigate Risks

Addressing the challenges and mitigating the risks associated with AI in construction requires a strategic and proactive approach. Here are some strategies that construction companies can adopt:

- *Develop a Data Strategy:* Establish a comprehensive data strategy that focuses on data quality, management, and governance. Invest in data cleansing and validation processes to ensure that AI models are trained on accurate and representative data.
- *Pilot Projects:* Begin with small-scale pilot projects to test AI applications in real-world construction scenarios. This allows for the identification of challenges and risks early on and provides an opportunity to fine-tune AI solutions.
- *Invest in Training:* Develop training programs to upskill the existing workforce.

The coordination of computerized reasoning (computer-based intelligence) in the development business has opened a domain of extraordinary potential outcomes, promising expanded productivity, improved well-being, and imaginative headways. Nonetheless, this change isn't absent of any and all difficulties and dangers that should be mindfully addressed to completely outfit the likely advantages of simulated intelligence while moderating its possible disadvantages. As the development area leaves on this innovative excursion, it should explore a complicated scene of moral, cultural, financial, and specialized contemplations to guarantee a dependable and practical man-made intelligence driven future. Man-made intelligence's implantation into development processes presents a scope of specialized difficulties that request cautious consideration. The development climate is intrinsically powerful, with complex connections between different parts, like materials, hardware, and human work. Creating AI frameworks that can actually comprehend, anticipate, and answer these complicated elements requires refined calculations and strong information inputs. Guaranteeing the dependability, exactness, and versatility of simulated intelligence models to oblige the different and consistently changing nature of development tasks remains as an imposing specialized challenge.

Moral contemplations come to the very front as simulated intelligence turns out to be profoundly entwined with development exercises. Issues of information protection, security, and straightforwardness emerge as man-made intelligence frameworks gather, investigate, and use immense measures of touchy data. Adjusting the requirement for information-driven experiences with the basic to safeguard individual freedoms and forestall likely abuse of information is a complex moral problem that should be addressed to cultivate public trust and trust in AI applications.

Also, the appearance of computer-based intelligence in development conveys possible cultural dangers, especially concerning position dislodging and monetary disparity. As simulated intelligence advancements computerize routine undertakings, concerns arise about the effect on the labor force, especially those involved in physical work or dull jobs. Guaranteeing a simply progress for laborers impacted by computer-based intelligence-driven changes is fundamental to abstain from intensifying existing cultural incongruities. Finding some kind of harmony between the advantages of man-made intelligence driven proficiency and the conservation of livelihoods represents a basic cultural test.

Monetary contemplations further highlight the requirement for a complete way to deal with dealing with simulated intelligence's difficulties and dangers. The forthright expenses related with executing man-made intelligence advancements, including foundation, preparing, and mix, can introduce boundaries for little and medium-sized ventures. Addressing these monetary obstacles and guaranteeing evenhanded admittance to computer-based intelligence-driven arrangements is urgent to forestall a situation where bigger, very much promoted organizations excessively receive the rewards, prompting a lopsided battleground and potential industry union.

To explore these multilayered difficulties and dangers, a comprehensive methodology is basic needs. Cooperation among partners—industry players, states, scientists, and common society—is fundamental to foster systems that guide capable man-made intelligence reception altogether. Vigorous administrative measures that address information security, moral rules, and guidelines for man-made intelligence in development can give an establishment to the business' development. Instructive drives and preparing programs should be laid out to engage the labor force with the abilities expected to team up successfully with simulated intelligence innovations, guaranteeing a smooth progress and limiting position removal.

All in all, while the reconciliation of computer-based intelligence in development guarantees groundbreaking progressions, it is joined by a range of difficulties and dangers that require proactive and smart moderation. Tending to the specialized intricacies, moral predicaments, cultural worries, and financial hindrances requires a purposeful work to encourage capable computer-based intelligence improvement and sending. Thus, the development business can outline a way toward a future where computer-based intelligence-driven developments benefit society on the loose while shielding against likely traps.

Data Quality and Bias

One of the significant challenges in leveraging AI in construction is ensuring the quality and reliability of the data used to train AI algorithms. Construction projects generate vast amounts of data, including design plans, sensor data, and historical project information (Saad et al., 2020). However, this data may be fragmented, inconsistent, or of varying quality, which can impact the accuracy and performance of AI models.

Moreover, bias in data is a critical concern. If AI systems are trained on biased data or programmed with biased rules, it can perpetuate and amplify existing societal biases. This could lead to discriminatory outcomes in decision-making processes, such as resource allocation, hiring practices, or project prioritization. Addressing these challenges requires careful data management practices and algorithmic transparency. It is essential to ensure data quality through robust data collection, cleaning, and validation processes. Additionally, regular audits of AI algorithms should be conducted to identify and mitigate biases. Implementing diverse and inclusive data sets and involving stakeholders in the design and validation of AI models can help address data quality and bias issues in AI-assisted construction.

Ensuring Safety and Security

The adoption of AI in construction also raises concerns about safety and security. AI systems rely on extensive data and complex algorithms to make decisions and recommendations. Ensuring the reliability and safety of these systems is crucial to prevent errors, accidents, or malicious attacks that could have significant consequences for construction projects and human lives. To address safety concerns, rigorous testing, validation, and verification processes should be implemented throughout the development and deployment of AI systems in construction. Collaborations between AI experts, construction professionals, and regulatory bodies can establish standards and best practices for AI safety in construction. Additionally, cybersecurity is a critical aspect to consider. AI systems in construction rely on networked devices, cloud storage, and data sharing, making them potential targets for cyberattacks (Naeem et al., 2022). Robust cybersecurity protocols, including encryption, access controls, and regular system audits, should be implemented to protect AI systems and the sensitive data they process (Shneiderman, 2020).

Ethical and Societal Implications of AI in Construction: Data Privacy and Security

The construction industry has not been exempted from the fast transformation that AI has brought about in many other industries. AI innovations are revolutionizing the construction sector by boosting output, efficiency, and security. But these innovations also raise a host of social and ethical problems that need to be properly considered. When using AI in construction, two significant concerns are data security and privacy. We will examine the ethical and sociological ramifications of AI in construction in this essay, with an emphasis on the difficulties and protections around data security and privacy.

The Role of AI in Construction

AI technologies in construction encompass a wide range of applications, including:

- *Project Management:* AI can assist in project planning, scheduling, and resource allocation, optimizing construction timelines and budgets.
- *Predictive Maintenance:* AI-driven predictive maintenance systems can monitor machinery and equipment, reducing downtime and ensuring worker safety.
- *Safety Monitoring:* AI-powered cameras and sensors can analyze job sites in real-time to detect potential safety hazards and accidents.
- *Quality Control:* AI can inspect construction materials and structures for defects, ensuring compliance with quality standards.
- *Design Optimization:* Generative design algorithms can create optimal building designs based on various parameters such as cost, energy efficiency, and aesthetics.
- *Supply Chain Management:* AI can optimize supply chain logistics, reducing delays and cost overruns.

While these AI applications promise numerous benefits, they also introduce ethical and societal considerations, particularly regarding the collection, use, and protection of data.

Data Privacy Concerns in AI-driven Construction

- *Data Collection:* Construction sites generate vast amounts of data, including project plans, blueprints, worker schedules, and equipment data. AI systems require access to this data to function effectively. However, collecting and storing this data can raise concerns about the privacy of individuals involved in the project, as well as sensitive project information.
- *Worker Privacy:* AI-powered surveillance systems may monitor workers to ensure safety and productivity. This raises questions about workers' rights and privacy, as constant monitoring can feel invasive.
- *Personal Information:* Construction projects often involve the exchange of personal information, such as employee records and contractor details. Ensuring the security and privacy of this data is crucial.
- *Ownership of Data:* Determining who owns the data generated by AI systems on construction sites can be complex. Construction companies, contractors, and AI solution providers must establish clear agreements regarding data ownership and usage rights.

Safeguards for Data Privacy and Security

To address these data privacy concerns, the construction industry can implement various safeguards:

- *Data Encryption:* All data transmitted and stored should be encrypted to protect it from unauthorized access.
- *Access Control:* Limit access to sensitive data to authorized personnel only and implement stringent authentication protocols.
- *Data Anonymization:* Anonymize data whenever you can to preserve people's privacy while yet allowing AI systems to operate properly.
- *Transparency:* Assure that all interested parties are informed about the data being gathered, how it will be used, and who will have access to it. Transparency fosters trust.
- *Data Retention Policies:* Establish clear policies for how long data will be retained and when it will be securely deleted.
- *Ethical Guidelines:* Develop and enforce ethical guidelines for the use of AI in construction, including worker monitoring and data handling.
- *Compliance with Regulations:* Ensure that AI systems and data practices comply with relevant data protection laws, such as GDPR in Europe and HIPAA in the United States.

Societal Implications

Beyond data privacy, the societal implications of AI in construction extend to various areas:

- *Job Displacement:* As AI and automation technologies become more prevalent, there is concern about job displacement in the construction industry (Qureshi et al., 2022). While AI can enhance productivity, it may also reduce the need for certain manual labor roles.

- *Skills Gap:* The implementation of AI requires a workforce with the skills to operate and maintain these technologies. Bridging the skills gap through training and education is essential.
- *Inequality:* The adoption of AI in construction may exacerbate socioeconomic inequalities if certain workers or regions are left behind in terms of access to training and opportunities.
- *Environmental Impact:* AI can optimize designs for energy efficiency and sustainability, but it also requires significant energy resources for its operation. Balancing these factors is crucial for minimizing the environmental impact of AI in construction.
- *Security:* While AI may improve security through danger identification and monitoring, there is a chance that an excessive dependence on technology could result in lax security procedures. The right balance between technology development and human attentiveness must be struck.

AI has the potential to alter the construction industry by enhancing quality, security, and efficiency. The moral and cultural implications, particularly those relating to data security and privacy, cannot be overlooked, though. To guarantee that AI is applied in a way that benefits society while upholding individual rights and privacy, construction businesses, regulators, and AI solution suppliers must collaborate to set strong data protection measures, ethical norms, and training programs. The construction sector can maximize AI's promise while minimizing its hazards by proactively addressing these worries (Robinson, 2020).

Protecting Sensitive Data

The use of AI in construction relies heavily on data, including sensitive information about individuals, projects, and organizations. This data may encompass proprietary designs, financial records, personal information, or intellectual property. As AI systems process and analyze this data, ensuring its privacy and protection becomes paramount.

Ensuring Cybersecurity

The integration of AI in construction introduces new challenges and risks in terms of cybersecurity. AI systems rely on networked devices, cloud storage, and data sharing, making them potential targets for cyberattacks. Malicious actors may seek to exploit vulnerabilities in AI systems or gain unauthorized access to sensitive information. To address these cybersecurity challenges, it is necessary to establish robust protocols and best practices. This includes implementing secure network architectures, employing encryption technologies, regularly updating software and systems to patch vulnerabilities, and conducting thorough system audits. It is also important to train personnel on cybersecurity practices and create awareness about potential threats and risks associated with AI in construction.

Collaboration between AI experts, construction professionals, and cybersecurity specialists can help develop comprehensive cybersecurity frameworks and guidelines specifically tailored for AI-assisted construction. Regulatory bodies can play a crucial role in establishing industry standards and enforcing compliance with cybersecurity practices to protect the integrity and security of AI systems in construction.

Data privacy refers to the control and protection of individuals' personal information. It is crucial to establish clear policies and frameworks for data collection, storage, and usage in AI-assisted construction. An ethical approach involves obtaining informed consent from individuals whose data is being collected, ensuring transparency about how the data will be used, and adhering to data protection regulations such as the GDPR. Furthermore, securing data against unauthorized access or breaches is essential. Implementing robust cybersecurity measures, including encryption, access controls, and regular system audits, helps safeguard sensitive data from cyber threats. It is also crucial to consider data anonymization techniques that protect privacy while allowing for the use of aggregated data for AI model training and analysis.

Ethical and Societal Implications of AI in Construction: Job Displacement and Automation

The rapid advancement of the digital era has led to the widespread replacement of human work. Some experts believe that such rapid expansion will have a substantial impact on human civilization and eventually lead to a significant automation of human labor. Nissim and Simon assert that automation and AI have the potential to damage companies that were built to be resilient. They continued by saying that in addition to defending their members' interests, unions have a moral duty to uphold everyone's moral values (Khogali & Mekid, 2023).

The integration of AI in the construction industry holds significant potential for enhancing productivity, efficiency, and safety. AI technologies can automate various tasks, optimize workflows, and enable data-driven decision-making. However, alongside these advancements, ethical and societal implications arise, particularly concerning job displacement and automation. This section explores the ethical and societal implications of AI in construction, with a specific focus on job displacement and the automation of tasks. The two key subheadings are given below:

The Impact on the Workforce

One of the primary concerns regarding the adoption of AI in construction is the potential displacement of human workers. AI technologies have the capability to automate repetitive, labor-intensive tasks, which could potentially reduce the demand for certain types of jobs within the construction industry. This displacement raises questions about the social and economic consequences for affected workers. Job displacement can lead to unemployment, income inequality, and societal disruptions. It is essential to approach the integration of AI in construction with a human-centric perspective, considering the potential impact on the workforce. This involves engaging in proactive strategies to address job displacement, such as developing transition programs, providing retraining opportunities, and facilitating the transition to new roles that complement AI technologies.

Moreover, it is crucial to prioritize the well-being of workers throughout this transition. This includes providing support systems, fostering a culture of continuous learning, and promoting the development of new job opportunities that leverage human skills alongside AI. By adopting these measures, ethical concerns related to job displacement can be mitigated, and the workforce can be empowered to adapt and thrive in the changing landscape of AI-assisted construction.

Reskilling and Job Creation

While job displacement is a concern, the integration of AI in construction also presents opportunities for reskilling and job creation. As AI takes over certain repetitive tasks, it frees up human workers to focus on higher-value activities that require creativity, critical thinking, problem-solving, and interpersonal skills. To harness these opportunities, it is crucial to invest in reskilling and upskilling programs for construction workers. By providing training in areas such as data analysis, AI implementation, project management, and collaboration with AI systems, workers can acquire the skills necessary to complement and collaborate with AI technologies. This approach ensures that workers remain relevant and able to leverage their unique human capabilities alongside AI. Additionally, the adoption of AI in construction can lead to the creation of new job roles and opportunities. These roles may include AI system trainers, AI project managers, data analysts, and technicians responsible for maintaining and optimizing AI systems. By fostering an ecosystem that supports innovation, entrepreneurship, and job creation, the integration of AI in construction can contribute to economic growth and the evolution of the industry.

Employers now have the choice to use automation to complete tasks that formerly needed human labor. Workers may fear that their jobs are under danger when firms look for ways to use automation technology. For both groups, this raises problems. Through their survey of 502 respondents from Bulgaria on their perspectives on work automation, the authors of *Automation Fears: Drivers and Solution* offered proof for this assertion. Due to increased worries about automation, the analysis revealed that personal solutions won out over commercial and societal ones. The poll found that people's concerns about job automation are influenced by their views and demography. Peer pressure, the job's automatability, beliefs about the dehumanizing effects of technology, and a person's self-perception of professionalism are the main elements associated with the fear of automation (Khogali & Mekid, 2023).

Ethical and Societal Implications of AI in Construction: Equity and Accessibility

The integration of AI in the construction industry has the potential to revolutionize processes, enhance efficiency, and drive innovation. However, as AI becomes increasingly prevalent, it is crucial to consider the ethical and societal implications, particularly in terms of equity and accessibility. This article explores the ethical and societal implications of AI in construction, with a specific focus on equity and accessibility. The two key subheadings are given below (Zallio & Clarkson, 2022).

Digital Divide

The adoption of AI in construction raises concerns about the potential exacerbation of the digital divide. The digital divide refers to the gap between individuals and communities who have access to technology and those who do not (Rasheed et al., 2021). It encompasses disparities in access to internet connectivity, hardware, software, and digital literacy skills. As AI technologies become integral to construction processes, there is a risk that smaller firms, individuals, or economically disadvantaged communities may be left behind due to limited access to AI tools and resources as shown in Fig 10.1. This could result in an uneven distribution of benefits and opportunities, further widening existing economic and social inequalities.

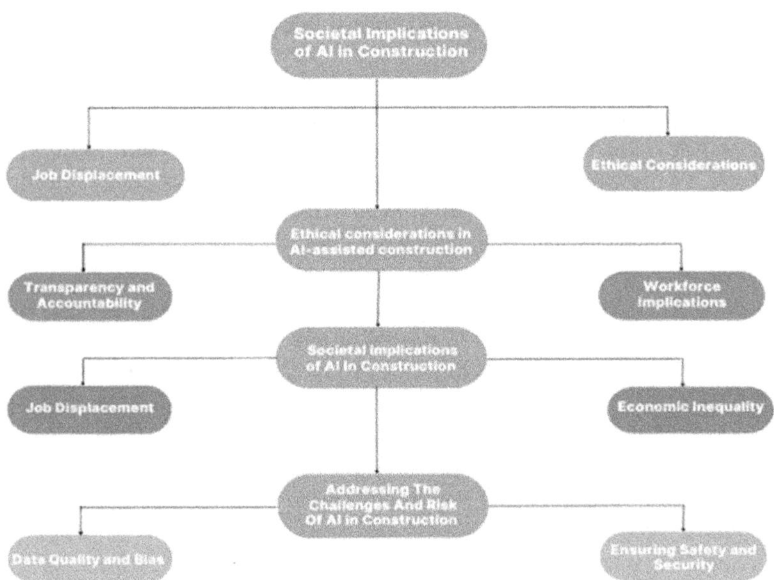

Figure 10.1: Flow chart for social implications of AI in construction.

To address the digital divide, it is essential to promote equity in access to AI technologies. This includes investing in infrastructure development to improve internet connectivity, providing affordable access to hardware and software, and offering training programs to enhance digital literacy skills. Collaborations between public and private sectors, as well as educational institutions, can help bridge the digital divide and ensure that AI benefits reach all segments of the construction industry.

Bias and Fairness

Another critical consideration in the ethical and societal implications of AI in construction is the potential for bias and unfairness. AI systems are trained on large datasets that may contain inherent biases, reflecting historical disparities or societal prejudices. If left unchecked, these biases can perpetuate discriminatory practices in decision-making processes related to resource allocation, project prioritization, or hiring practices.

To address bias and ensure fairness, it is crucial to adopt a proactive approach. This includes conducting regular audits of AI algorithms and datasets to identify and mitigate biases. The development and use of diverse and inclusive datasets can help reduce bias and ensure that AI systems are representative of diverse perspectives and experiences. Moreover, involving diverse stakeholders, including underrepresented groups, in the design, development, and validation of AI systems can help identify and rectify potential biases as shown in Table 10.1. Transparency and accountability in AI systems are also essential, ensuring that the decision-making processes of an AI algorithms are explainable and auditable.

Case Study

"Research and Practice of AI Ethics: A Case Study Approach Juxtaposing Academic Discourse with Organisational Reality". This case study delves into the ethical implications surrounding the utilization of Big Data and AI technologies, collectively referred to as BD+AI, through an empirical investigation. The research systematically categorizes existing literature on this topic and employs a multi-case study approach to explore practical ethical dilemmas in various contexts. By employing qualitative tools, the study analyzes data from ten specific case studies spanning diverse domains. The aim is to amalgamate distinct ethical concerns, as documented in the literature, into cohesive clusters, subsequently comparing them with the established classification. The findings of this study unveil intriguing insights. Despite the wide array of social domains, fields, and applications in which AI and BD are applied, a degree of overlap and correlation emerges in terms of the ethical apprehensions faced by different organizations. This points to a shared thread of ethical challenges that traverse diverse sectors. The study's focus on practical, real-world scenarios provides a nuanced comprehension of the ethical landscape inherent in the realm of AI+BD.

Table 10.1: Comparative analysis for social implications of AI in construction

Topic	Ethical and societal implications	Ethical considerations	Data privacy and security	Accessibility
Job Displacement	Potential loss of construction jobs due to automation.	Fair treatment of displaced workers.	Secure storage and handling of worker data.	Ensuring AI doesn't create barriers for individuals with disabilities.
Safety	AI errors or malfunctions may lead to safety risks.	Ensuring AI systems undergo rigorous testing and validation.	Protecting sensitive project and worker information.	Providing accessible safety information for all workers.
Bias and Fairness	Biased AI algorithms may perpetuate social inequalities.	Developing and using unbiased AI models.	Safeguarding data from unauthorized access.	Making AI interfaces and tools intuitive for all users.
Data Ownership and Use	Construction data may be exploited or misused.	Establishing clear data ownership rights.	Implementing encryption and access controls for data.	Ensuring AI tools are available to all stakeholders.
Environmental Impact	AI adoption may impact the environment through increased electronic waste.	Considering sustainable AI infrastructure.	Safeguarding sensitive environmental data.	Making AI tools available in eco-friendly formats.
Project Costs and Affordability	AI integration costs may increase construction expenses.	Balancing AI adoption to benefit the wider community.	Protecting financial data and transactions.	Providing cost-effective AI solutions for all stakeholders.
Accountability	Assigning responsibility for AI-driven decisions.	Developing transparent AI systems.	Ensuring compliance with data protection regulations.	Offering support for those unfamiliar with AI technologies.

This enhanced understanding of ethical considerations in AI+BD assumes pivotal significance. It serves as a compass to navigate the myriad proposed solutions aimed at addressing these ethical quandaries. By pinpointing the specific ethical predicaments that often dominate scholarly discourse, this research enables a more targeted and effective approach to mitigating these concerns. By bridging the gap between theoretical propositions and practical manifestations, the study contributes to the creation of comprehensive frameworks that can address ethical dilemmas in a pragmatic manner. To achieve these outcomes, the study employs a multifaceted methodology. It first synthesizes existing literature, creating a taxonomy of prevalent ethical concerns associated with BD+AI technologies. Subsequently, it conducts an in-depth exploration of real-world cases, each situated within a distinct domain. This multi-case approach enriches the study by grounding it in practical experiences and shedding light on context-specific nuances.

The synthesis of empirical data and theoretical underpinnings culminates in a more holistic perspective on ethical considerations in the BD+AI landscape. This research underscores the interdisciplinary nature of the field, highlighting the convergence of technology, ethics, and societal impact. By examining real cases, the study transcends the realm of speculation, offering tangible insights into the complex interplay between technology and ethics

In conclusion, this study underscores the critical importance of comprehending and addressing ethical implications in the intersection of Big Data and AI. Through an empirical investigation that melds theoretical frameworks with real-world scenarios, the research reveals common threads of ethical concerns across diverse domains. By shedding light on these shared challenges, the study provides a roadmap for more targeted and effective ethical interventions. Ultimately, this work advances the discourse on ethical considerations in the realm of AI+BD, facilitating a more responsible and accountable deployment of these transformative technologies.

References

Abbasi, M.F., Bilal, M. and Rasheed, K. (2022). Role of Human Intuition in AI Aided Managerial Decision Making: A Review. *2022 International Conference on Decision Aid Sciences and Applications, DASA 2022*. https://doi.org/10.1109/DASA54658.2022.9765153.

Abioye, S.O., Oyedele, L.O., Akanbi, L., Ajayi, A., Davila Delgado, J.M., Bilal, M., Akinade, O.O. and Ahmed, A. (2021). Artificial intelligence in the construction industry: A review of present status, opportunities and future challenges. *Journal of Building Engineering*, 44. https://doi.org/10.1016/j.jobe.2021.103299.

Aldoseri, A., Al-Khalifa, K.N. and Hamouda, A.M. (2023). Re-thinking data strategy and integration for artificial intelligence: Concepts, opportunities, and challenges. *Applied Sciences*, 13(12). https://doi.org/10.3390/app13127082

Ammad, S., Alaloul, W.S., Saad, S. and Qureshi, A.H. (2021). Personal protective equipment (PPE) usage in construction projects: A systematic review and smart PLS approach. *Ain Shams Engineering Journal*, 12(4), 3495–3507. https://doi.org/https://doi.org/10.1016/j.asej.2021.04.001.

Ammad, S., Saad, S., Bashir, M.T., Qureshi, A.H., Altaf, M. and Rasheed, K. (2021). Construction Accidents via Integrating Building Information Modelling (BIM) with Emerging Digital Technologies: A Review. *2021 3rd International Sustainability and Resilience Conference: Climate Change*. https://doi.org/10.1109/IEEECONF53624.2021.9668066.

Chen, L., Chen, Z., Zhang, Y., Liu, Y., Osman, A.I., Farghali, M., Hua, J., Al-Fatesh, A., Ihara, I., Rooney, D.W. and Yap, P.-S. (2023). Artificial intelligence-based solutions for climate change: A review. *Environmental Chemistry Letters*, 21(5), 2525–2557. https://doi.org/10.1007/s10311-023-01617-y.

de Soto, B.G., Georgescu, A., Mantha, B., Turk, Ž. and Maciel, A. (2020). *Construction Cybersecurity and Critical Infrastructure Protection: Significance, Overlaps, and Proposed Action Plan*. https://doi.10.20944preprints202005.0213.v1.

Dwivedi, Y.K., Hughes, L., Ismagilova, E., Aarts, G., Coombs, C., Crick, T., Duan, Y., Dwivedi, R., Edwards, J. and Eirug, A. (2021). Artificial Intelligence (AI): Multidisciplinary perspectives on emerging challenges, opportunities, and agenda for research, practice and policy. *International Journal of Information Management*, 57, 101994.

Fountaine, T., McCarthy, B. and Saleh, T. (2019). Building the AI-powered organization. *Harvard Business Review*, 97(4), 62–73.

George, G., McGahan, A.M. and Prabhu, J. (2012). Innovation for inclusive growth: Towards a theoretical framework and a research agenda. *Journal of Management Studies*, 49(4), 661–683.

Hamouda, Abdel Magid, Abdulaziz Aldoseri and Khalifa N. Al-Khalifa (2023). Re-thinking data strategy and integration for artificial intelligence: Concepts, Opportunities, and Challenges. *Applied Sciences*, 13(12), 7082. MDPI.

Issa, H., Sun, T. and Vasarhelyi, M.A. (2016). Research ideas for artificial intelligence in auditing: The formalization of audit and workforce supplementation. *Journal of Emerging Technologies in Accounting*, 13(2), 1–20.

Jiang, B., Khan, R.R.A., Vian, A. and Cheng, Z. (2015). BIM implementation in China: A case study approach. *ICITMI*, 1106–1113. https://doi.org/10.2991/icitmi-15.2015.186.

Khogali, H.O. and Mekid, S. (2023). The blended future of automation and AI: Examining some long-term societal and ethical impact features. *Technology in Society*, 73, 102232. https://doi.org/https://doi.org/10.1016/j.techsoc.2023.102232.

Kiriiri, G.K., Njogu, P.M. and Mwangi, A.N. (2020). Exploring different approaches to improve the success of drug discovery and development projects: A review. *Future Journal of Pharmaceutical Sciences*, 6(1), 27. https://doi.org/10.1186/s43094-020-00047-9.

Madni, A.M. and Jackson, S. (2009). Towards a conceptual framework for resilience engineering. *IEEE Systems Journal*, 3(2), 181–191. https://doi.org/10.1109/JSYST.2009.2017397.

Mitchell, A.G. (1998). *Strategic Training Partnerships between the State and Enterprises*. International Labour Organization.

Mohsen Alawag, A., Salah Alaloul, W., Liew, M.S., Ali Musarat, M., Baarimah, A.O., Saad, S. and Ammad, S. (2023). Critical success factors influencing total quality management in industrialized building system: A case of Malaysian Construction Industry. *Ain Shams Engineering Journal*, 14(2), 101877. https://doi.org/https://doi.org/10.1016/j.asej.2022.101877.

Moniz, A.B. and Krings, B.-J. (2016). Robots working with humans or humans working with robots? Searching for social dimensions in new human-robot interaction in industry. *Societies*, 6(3), 23.

Mourtzis, D., Angelopoulos, J. and Panopoulos, N. (2022). A literature review of the challenges and opportunities of the transition from Industry 4.0 to Society 5.0. *Energies*, 15(17), 6276.

Naeem, U.H., Bilal, M., Syed, F., Rasheed, K. and Saad, S. (2022). A Multilayer Encryption Model to Protect Healthcare Data in Cloud Environment. *2022 International Conference on Data Analytics for Business and Industry, ICDABI 2022*. https://doi.org/10.1109/ICDABI56818.2022.10041708.

Osterman, P., Locke, R.M. and Piore, M.J. (2002). *Working in America: A Blueprint for the New Labor Market*. MIT Press.

Pan, Y. and Zhang, L. (2021). Roles of artificial intelligence in construction engineering and management: A critical review and future trends. *Automation in Construction*, 122, 103517. https://doi.org/https://doi.org/10.1016/j.autcon.2020.103517.

Panth, B. and Maclean, R. (2020). Introductory overview: Anticipating and preparing for emerging skills and jobs—Issues, concerns, and prospects. *In: Education in the Asia-Pacific Region* (Vol. 55). https://doi.org/10.1007/978-981-15-7018-6_1

Perifanis, N.-A. and Kitsios, F. (2023). Investigating the influence of artificial intelligence on business value in the digital era of strategy: A literature review. *Information*, 14(2). https://doi.org/10.3390/info14020085.

Qureshi, A.H., Alaloul, W.S., Manzoor, B., Musarat, M.A., Saad, S. and Ammad, S. (2020). Implications of Machine Learning Integrated Technologies for Construction Progress Detection under Industry 4.0 (IR 4.0). *2020 Second International Sustainability and Resilience Conference: Technology and Innovation in Building Designs (51154)*, 1–6. https://doi.org/10.1109/IEEECONF51154.2020.9319974.

Qureshi, A.H., Alaloul, W.S., Wing, W.K., Saad, S., Alzubi, K.M. and Musarat, M.A. (2022). Factors affecting the implementation of automated progress monitoring of rebar using vision-based technologies. *Construction Innovation* (ahead-of-print). https://doi.org/10.1108/CI-04-2022-0076.

Rasheed, K., Saad, S., Shahzad, L., Ammad, S., Ali, A. and Badshah, I. (2021). Application of Gesture Data Recognition in a Human-Interactive Leap Motion Sensor Chair. *2021 International Conference on Data Analytics for Business and Industry, ICDABI 2021*. https://doi.org/10.1109/ICDABI53623.2021.9655819

Robinson, S.C. (2020). Trust, transparency, and openness: How inclusion of cultural values shapes Nordic national public policy strategies for artificial intelligence (AI). *Technology in Society*, 63, 101421.

Saad, S., Alaloul, W.S., Ammad, S., Qureshi, A.H., Altaf, M. and Rasheed, K. (2020). Design Phase Carbon Emission Prediction Using a Visual Programming Technique. *2020 2nd International Sustainability and Resilience Conference: Technology and Innovation in Building Designs*. https://doi.org/10.1109/IEEECONF51154.2020.9319978.

Saad, S., Alaloul, W.S., Ammad, S. and Qureshi, A.H. (2022). A qualitative conceptual framework to tackle skill shortages in offsite construction industry: A scientometric approach. *Engineering, Construction and Architectural Management*, 29(10), 3917–3947. https://doi.org/10.1108/ECAM-04-2021-0287.

Shneiderman, B. (2020). Bridging the gap between ethics and practice: Guidelines for reliable, safe, and trustworthy human-centered AI systems. *ACM Transactions on Interactive Intelligent Systems (TIIS)*, 10(4), 1–31.

Teng, S.Y., Touš, M., Leong, W.D., How, B.S., Lam, H.L. and Máša, V. (2021). Recent advances on industrial data-driven energy savings: Digital twins and infrastructures. *Renewable and Sustainable Energy Reviews*, 135, 110208. https://doi.org/https://doi.org/10.1016/j.rser.2020.110208.

Weber-Lewerenz, B. (2021). Corporate digital responsibility (CDR) in construction engineering—Ethical guidelines for the application of digital transformation and artificial intelligence (AI) in user practice. *SN Applied Sciences*, 3, 1–25.

Xia, L., Semirumi, D.T. and Rezaei, R. (2023). A thorough examination of smart city applications: Exploring challenges and solutions throughout the life cycle with emphasis on safeguarding citizen privacy. *Sustainable Cities and Society*, 98, 104771.

Yermack, D. (2010). Shareholder voting and corporate governance. *Annu. Rev. Financ. Econ.*, 2(1), 103–125.

Yigitcanlar, T. and Cugurullo, F. (2020). The sustainability of Artificial Intelligence: An urbanistic viewpoint from the lens of smart and sustainable cities. *Sustainability*, 12(20). https://doi.org/10.3390/su12208548.

Zallio, M. and Clarkson, P.J. (2022). Designing the metaverse: A study on inclusion, diversity, equity, accessibility and safety for digital immersive environments. *Telematics and Informatics*, 75, 101909. https://doi.org/https://doi.org/10.1016/j.tele.2022.101909.

CHAPTER

11

AI Future Perspectives and Trends in Construction

Zawar Ali, Syed Saad, Kumeel Rasheed and Syed Ammad

Overview

We are going deeply into the fascinating world of artificial intelligence (AI) and all the fantastic possibilities that lie ahead in this chapter. Imagine a world where AI goes beyond simply automating jobs and instead collaborates with humans to develop novel concepts and creative solutions. We're starting a trip to learn how AI is developing into a more creative force that can help us solve complex issues that require human-like reasoning. This journey gives us a compelling look into a time where AI isn't just a tool, but a true innovation partner who piques our curiosity and alters the fundamental way we approach jobs. Nevertheless, there is a canvas of essential questions that must be examined amid all the hoopla. Consider the possibility that AI could have emotions or perhaps awareness. Our investigation also includes moral issues, like whether or not we should treat these robots as if they have some 'life' and how we ought to interact with them morally. This riddle challenges us to consider our humanity and how AI may become deeply entwined with our lives (Himeur et al., 2023).

The chapter also illuminates the broad impact of AI on numerous facets of our society. We are exposed to a wave of transformation, ranging from the healthcare industry, where AI transforms patient care, to the corporate world, where AI-driven decisions create waves. We'll see how AI is already making an impact through real-world instances. Then we'll imagine the tremendous changes it could bring about, potentially changing labor markets and the fundamental dynamics of our society. And now for a beautiful twist: AI's strike into the world of creativity! Be prepared for discussions about how AI is sparking creativity and producing stories, music, and art. Imagine having a robot pal who can also paint or tell stories. This chapter demonstrates that AI's use goes beyond simple calculations and logic to embrace creativity and push the envelope of what is possible. As our investigation draws closer, an image of AI acting as a catalyst for improved thinking emerges. Imagine it as a wise buddy that expands our comprehension in novel ways. We will investigate how AI may make learning an exciting trip and allow us to explore unexplored intellectual realms through this lens. Prepare for a journey into the future where AI redefines our reality, igniting our creative powers and lifting us to higher levels of thinking ability (Bhowmik et al., 2022).

Emerging Technologies in AI and Material Science

The global pandemic's effects significantly impacted the previous year's significant period of technical advancement. A new way of life was introduced by Covid-19, forcing us to become accustomed to distant work and accelerating the digital transformation process. It's critical to examine the trends and technologies already influencing the landscape of 2021 and what potential opportunities might be on the horizon as we go into the ensuing year of exceptional technological advancement. The current year is in the early phases of what is frequently referred to as "Industry 4.0", a revolutionary period of technological transition marked by the merging of physical and digital systems. Healthcare, transportation, energy, construction, finance, commerce, and security are just a few of the industries on which this disruptive technological upswing is impacting. Automation, genomics, precision medicine, deep learning, 5G, the Internet of Things (IoT), material science, and autonomous transportation are just a few emerging technologies powering Industry 4.0. It's important to remember how frequently these applications cross into more general technological domains, weaving a tapestry of invention that ushers in exponential change (Ashebir et al., 2007).

The field of material science stands out among these revolutionary forces. The development of society has been substantially shaped by advances in materials over millennia of human history, from the Paleolithic Age to the upcoming fourth industrial revolution. Fundamentally, material science aims to understand the complex connections between various materials' composition, attributes, production methods, and uses. With the advent of Industry 4.0, the search for novel materials has taken on greater relevance and evolved into a driver for societal advancement. In material science, a sizable body of information has accumulated over millennia. However, ingesting and understanding the constantly growing volumes of material and data published each day has proven difficult due to the intrinsic cognitive limits of humans. Due to this restriction, only a tiny portion of the data can be analyzed inside specific subfields. The current method of material research mainly relies on a "trial-and-error" approach, supported by a variety of experimental endeavors informed by empirical knowledge, complemented by a scattering of computer simulations. However, this method requires significant time, money, and human resources. Databases provide a vast pool of untapped material information; new approaches are needed to utilize this resource fully (Lundvall et al., 2009). A new era of material science opportunities has arrived because of AI. Since its inception over 60 years ago, AI has developed from simple perceptions to sophisticated multilayer neural networks, building a robust algorithmic framework and potent hardware base. Some AI systems have even beaten human champions in games like Go, chess, and other contests. The extraordinary data mining capabilities of AI have captured the attention of the community of material scientists, sparking the development of "materials informatics"—an interdisciplinary field that combines material science and AI methods. This symbiotic relationship has excellent promise since it provides researchers with a fresh technique to anticipate material properties, predict hidden correlations between variables, direct chemical synthesis pathways, optimize process parameters, and improve current methods for material characterization.

The emergence of Industry 4.0 and the integration of AI into material science are two colossal forces coming together to generate a powerful narrative of innovation and change. These currents are propelling us towards unexplored regions of development, from the disruption brought on by Covid-19 to the emergence of Industry 4.0 and the catalytic role of AI in material science. It is a story of human creativity blending with technological prowess, pointing us toward a future in which the union of AI and material science may lead to achievements that advance society into previously unimaginable spheres of growth and prosperity (Fava et al., 2015).

Examples

AI enables scientists to quickly find and invent novel materials, revolutionizing the research rate using vast materials databases and cutting-edge machine learning algorithms. AI is revealing insights that defy human comprehension and directing the synthesis of customized metals, polymers, semiconductors, and chemicals with custom-engineered features through automated data analysis and the discovery of previously concealed structure-property connections (López, 2023). Reducing the need for time-consuming experimental trials is the ability of neural networks trained on massive databases to predict material compositions and qualities. By autonomously interpreting imaging and spectroscopic outputs, AI optimizes materials characterization equipment like electron microscopes, X-ray spectrometers, and synchrotrons, increasing comprehension at the atomic level. AI can also be used to synthesize knowledge from many fields through federated materials graph networks, leading to the discovery of novel material classes like van der Waals hetero-structures. In self-driving labs, the combination of AI and robotics enables automated testing and closed-loop optimization, realizing the potential for on-demand advanced manufacture of custom nanocomposite alloys, ceramics, and biomimetics at scales beyond human capability.

Advances in Machine Learning (ML)

Indeed, the quick development of ML techniques has been a critical stimulant for the development of AI. In this area, the introduction of deep learning is a paradigm shifting subfield that allows computers to understand complex patterns and carry out tasks via example-based learning without explicit programming. These deep learning algorithms are ushered in an era of unparalleled innovation, sparking paradigm shifts in various industries, including transportation and healthcare. Several cutting-edge deep learning technologies are highlighted on stage, capturing both interest and imagination. The game theoretic dance between two neural networks, a generator, and a discriminator, is orchestrated by the Generative Adversarial Networks (GANs), one of these stars (Pan et al., 2019). In this symphony of competition, the generator makes an effort to generate new data that closely resembles the statistics of the initial training set to trick the discriminator, which then improves its ability to recognize authenticity. GANs are transforming the creative world, enhancing photorealistic picture synthesis and the aesthetics of video games. GANs have been enthusiastically explored to include molecule production, which promises to speed up drug development through clever computational manipulation (Tee & Ouyang, 2018).

Reinforcement Learning emerges, taking cues from behaviorist psychology and allowing computers to learn complicated behaviors by interacting with their surroundings. The story of AlphaGo and Alpha Zero, created by DeepMind's creativity, exemplifies this power as they superhumanly defeated the complex strategies of the ancient board games of Go and chess without the aid of human skill. These principles are being applied to robotics, process control, game dynamics, and a variety of sequential decision-making puzzles in the developing chapters, creating opportunities for unparalleled automation and mastery (Arinez et al., 2020). In the meantime, the spotlight is widened to reveal the genius of the Transformers, a genealogy that includes characters like BERT, GPT-3, and Megatron. Based on the ground-breaking Transformer architecture, these models do away with the conventional reliance on recurrence or convolution in favor of attention processes. By doing this, they have effectively overcome translation, question-answering, and summarization obstacles and reinvented the field of natural language processing (NLP). Behemoths like GPT-3 have stunned the world with their amazingly human-like writing abilities, slowly blurring the borders between human and machine composition with their grand-scale manifestation. Transformers stand as sentinels at the threshold of a textual renaissance, with the seismic potential to transform how we comprehend and produce language.

The finest works of deep learning are used to paint the canvas of invention in this age of computational wonders. These technologies weave a narrative of AI evolution that enthralls both enthusiasts and academics, providing a glimpse into the seemingly limitless possibilities. From GANs that conjure novel realities to Reinforcement Learning's dance with the environment and Transformers that redefine linguistic horizons (Alimi & Meijboom, 2021).

Materials Discovery through Machine Learning

The accelerated discovery of ground-breaking materials is a field bursting with great potential in the constantly changing world of ML developments. ML introduces a dynamic set of tools that promise to orchestrate the discovery of novel materials within the digital realm (Hippalgaonkar et al., 2023), at a fraction of the time and cost typically associated with physical experimentation, in a field where traditional experimental methods frequently tread cautiously. Notable initiatives are emerging in this developing field, revealing the profound influence of machine learning on the field of material discovery. The field of Constitutional Materials Discovery is at the vanguard, and Anthropic's innovative minds have created Claude, a vast language model with the astonishing ability to suggest novel metal alloys and compositions. This technological marvel carefully examines millions of previously collected material data points to offer designs that align with desired qualities. Experimental evidence supporting the validity of Claude's ideas shows an exceptional success rate and points toward a day when material synthesis synchronizes flawlessly with computational accuracy (Wang et al., 2022).

2D Materials Discovery has undergone a seismic upheaval as well. Citrine Informatics' visionaries have used the power of ML algorithms to evaluate potential 2D materials for characteristics that could lead to electron band gaps suitable for electrical applications (Himanen et al., 2019). Their predictions have helped to reveal a fresh class of 2D semiconductors that were cleverly created from the trifecta of boron, nitrogen, and carbon, which is evidence of the effectiveness of this approach. The brilliant minds of DeepMind use their innovation, Alchemy, to engineer a significant change in Chemical Synthesis Planning as well. Alchemy takes on the role of mastering retrosynthesis planning by fusing the power of graph neural networks and reinforcement learning and offers multi-step chemical reaction ideas as a result. Alchemy has the potential to revitalize the pharmaceutical and fine chemical synthesis field, pushing it to previously unimaginable levels of efficiency and ingenuity (Naser, 2019).

The story does not end there, though. The phenomena of Accelerated Molecular Dynamics, a ground-breaking realization of ML's prowess in accelerating computationally intensive quantum mechanical computations within molecular dynamics simulations, is the result of the collaboration of the brains from Anthropic, Google, and DeepMind. This synergy accelerates modeling material properties and behaviors to an atomic resolution by approximating potential energy surfaces, giving a vantage point that reveals the concealed subtleties of matter (Tătaru et al., 2021). ML in material discovery stands out as an example of creativity in the middle of innovation and exploration. The journey encapsulates a testament to the transformative capabilities of ML in ushering in a new age of material creation, poised to shape the trajectory of industries and redefine the boundaries of scientific discovery with Constitutional Materials Discovery, 2D Materials Exploration, Chemical Synthesis Planning, and Accelerated Molecular Dynamics as the focal points.

Future Perspectives and Trends in AI-assisted Construction

The construction industry is undergoing a significant transformation as it persistently adopts the convergence of cutting-edge technologies, including AI, robotics, automation, and data analytics. The industry is going on a dramatic trajectory powered by innovation and departing from conventional labor-intensive ways, altering its core. The project lifecycle is transformed by AI, which brings remarkable efficiency from the beginning of conceptual planning through the precise orchestration of construction and subsequent facilities management. Generative design enhances the planning landscape by utilizing large datasets to quickly browse possibilities, and digital twin simulations reveal virtual validations of projects. Intelligent scheduling that optimizes timeframes and resource allocation based on past trends is the next movement in the symphony (Ashebir et al., 2007). Drones connected to AI and equipped with computer vision autonomously supervise construction sites, assuring worker safety and tracking progress. This dynamic supports the optimization of operations. A crucial role is played by the field of robotics, which frees up human hands from dangerous jobs like welding and the delicate choreography of 3D-printed formwork. AI-enhanced exoskeletons allow workers to carry larger loads while fending off tiredness. Tools that smoothly span the gap between local and distant teams and promote seamless communication result from the combination of AI and internet of things (IoT). Through defect identification, computer vision drives automated quality control and maintenance, while predictive algorithms are integrated into the fabric of supply-chain management to reduce waste and avoid delays (Niu et al., 2020). AI empowers autonomous heavy gear, enabling faster completion of tasks like carefully excavating exact foundations and 3D-printing tunnels. AI maps the way towards intelligent retrofitting and repairs, responding to changing needs over time by building on digital records of construction and material usage.

The industry is poised for unmatched changes as the convergence of AI, robots, automation, drones, and analytics approaches. Using advanced building methods supported by data-driven intelligent systems will allow for fine-tuning quality, resilience, and sustainability criteria while simultaneously reducing costs and project durations. The fabric that forms the spine of the building infrastructure is prepared to meet social needs with a higher degree of effectiveness and refinement.

AI for Precision Planning and Design

Building project planning and design are being revolutionized by AI. Extensive collections of 3D building models, simulations, and methods enable AI to demonstrate true creativity in various applications. Deep learning algorithms analyze previously honored projects' architectural designs and structural arrangements to make creative design recommendations. AI can enhance building functionality while carefully considering cost, material, and code compliance considerations when gently incorporated with feedback from human architects (Bag et al., 2021).

AI is also automating compliance checks for zoning approvals, streamlining these procedures. AI can swiftly discover potential difficulties by carefully reviewing design schematics and comparing them with zoning laws, ordinances, and land-use patterns—speeding up review timeframes. The workflow in urban planning is drastically altered as a result. AI significantly improves the precision with clever scheduling applications. ML algorithms create optimized timelines that account for work dependencies, staff availability, and unforeseen delays to assist projects in staying on budget and schedule by analyzing enormous volumes of schedule performance data from prior similar projects.

AI and virtual/augmented reality are used in design evaluations, creating new opportunities for international cooperation. Global stakeholders may easily coordinate while viewing designs in immersive 3D. Most importantly, AI's digital twin simulations can uncover operating trends even before construction is started by fusing 3D building/site models with IoT sensor data from live structures (S. Ammad et al., 2021). Inefficiencies are prevented and fixed, ensuring infrastructure continues serving communities as intended for many years. Across the AEC (architecture, engineering, and construction) industry, AI is orchestrating a symphony of precise planning and design.

Robotics and Automation on Construction Sites

Construction site jobs that are dangerous and physically demanding are changing thanks to robotics and automation technology made possible by AI. Heavy machinery like excavators can now work independently and with a level of accuracy far exceeding that of humans because of advances in machine vision and sensors (Rasheed, Saad et al., 2021). AI directs excavators to dig accurately by digital project plans, reducing incorrect digging and guaranteeing that the ground is adequately prepared for foundations. Similarly, robots that lay bricks and blocks at extraordinary speeds can construct walls without stopping.

Drones used for construction that have high-resolution cameras and other sensors are also enhancing safety oversight (Choi et al., 2023). Large projects may be monitored by AI-powered inspection drones, which can immediately spot even the most minor problems in difficult-to-reach areas that could harm personnel. Drones and 3D printing robots powered by AI and controlled through the cloud assist in quickly constructing temporary shelters on-site with the minor human intervention in disaster response scenarios (Tătaru et al., 2021).

AI-powered Smart Construction Management

AI is changing the world through data-driven analytics and digital tools. AI-driven cloud-based technologies easily integrate various project data streams in real-time, enabling previously unheard-of advancements in site operations by automatically matching real-world site photographs with 3D models and intuitively highlighting any deviations that might influence the timeline, computer vision and automation to improve progress tracking. With real-time RFID (radio-frequency identification) and sensor data capturing material movement and enabling predictive demand forecasting for smooth, just-in-time delivery (Rasheed, Shahzad et al., 2021), AI significantly impacts managing supply chain management. AI analyses past crew productivity trends against job demands to optimize workforce allocation, which then intelligently assigns people based on their abilities and location on-site for maximum effectiveness. Sensor networks vigilantly monitor sites for noise, vibration, and other characteristics. AI-driven anomaly detection strengthens safety (Himeur et al., 2023). ML swiftly identifies possible mishaps and automates alerts for quick managerial intervention. By combining augmented reality (AR) and visual reality (VR), stakeholders can easily coordinate through immersive 3D issue reviews, raising the bar for collaboration. In conclusion, AI emerges as the critical driver behind unheard-of improvements in construction management, integrating data-driven digital solutions to unify oversight, resource planning, and operations, ushering in cost savings and on-time project deliveries.

Opportunities for Further Research and Development

The search for even more sophisticated and useful AI forms continues to be unstoppable in the constantly changing field of AI, which has made significant progress. The identification of many crucial domains ready for continued investigation serves to highlight this path. To overcome enduring difficulties, the insatiable appetite for more advanced data and computing capabilities emerges as a key factor. These difficulties are caused by the lack of sufficient training instances, which makes it essential to compile and annotate a variety of datasets to promote better system generalization. With increased cloud and edge computing capabilities, which are expected to support the construction of bigger, more effective models, the horizon opens out even further. Algorithmic research is a crucial pillar in this complicated tapestry, delivering cutting-edge methods in deep learning, reinforcement learning, transfer learning, self-supervised learning, and more to enhance learning from scenarios with little to no data. Hybrid models, which combine many methodologies, hold the alluring potential to simplify jobs requiring a variety of skill sets (Lundvall et al., 2009). Model inspection, edge case validation, and formal verification strategies all act as sentinels to strengthen system reliability and prevent biases or mistakes from spreading unintentionally. The quest for AI's ultimate potential entails the creation of unified models ready for multi-domain, multi-task learning, supported by cognitive processes that resemble those of humans. This path may involve ground-breaking designs that seamlessly integrate neural networks with theories of causal learning and symbolic reasoning. The trend simultaneously shifts towards the symbiotic coexistence of humans and AI, imagining AI partners skilled at interpreting human nuances, explaining their reasoning, and cooperatively enhancing human skills (Tătaru et al., 2021). The underlying belief that concentrated efforts across several domains—algorithms, data, trust, general intelligence, and human aspects—hold the key to unlocking the full, secure, and egalitarian potential of AI is what unites this diverse research. This confirms that sustained and major research efforts are required in the quest to fully exploit AI's safe and equitable benefits. The appeal for continuous, deliberate research is thus once again heard.

Improving Model Generalization

While AI has made tremendous progress in producing outstanding results under confined boundaries, the ongoing challenge is increasing AI's ability to generalize and adapt across broader contexts. This vital endeavor demands a deeper dive into improving machine learning methods to support flexibility. In the current environment, neural networks perform well when restricted to the training data but struggle in uncharted territory. Researchers are actively investigating cutting-edge techniques, such as synthetic data augmentation and subjecting models to a wide range of testing situations to uncover the actual depths of these models' comprehension capacities (Wang et al., 2022). The future of AI systems is one in which they continuously learn from user interactions and fresh data, acquiring pertinent skills and knowledge throughout their operational lifetimes. Models can adapt and improve themselves in real-world settings thanks to the self-supervised and reinforcement learning approaches that developers are developing. The effectiveness of pre-training strategies is further demonstrated by their capacity to transfer information between related areas using sizable unlabeled datasets. Self-supervised methods provide a promising way to use surrogate prediction tasks to bootstrap fundamental ideas and linguistic patterns, leading to flexible learning (Tătaru et al., 2021). This learning could be applied easily as models are refined with fewer labelled instances in other tasks. Such an approach can enable AI with robust and flexible understanding capabilities, overcoming the limitations of current rigid training paradigms by merging diverse research areas like evaluating edge situations, unsupervised pre-training, domain adaption, and online updates. The practical application of AI technology in many real-world scenarios is expected to improve due to this development.

Understanding Model Decisions

In the evolving landscape of AI, particularly in critical decision-making domains, comprehending the rationale behind AI's choices must be balanced. A focused research effort is directed towards enhancing interpretability and generating explanations, a cornerstone for these technologies' responsible and reliable application. Ongoing endeavors involve developing techniques such as counterfactual explanations, illuminating how slight input variations could alter model predictions, and assessing predictive sensitivity. Simultaneously, exploring interactive debugging tools allows human interaction with models, enabling queries and visualizations of internal processes for deeper insights into decision logic. Another vibrant area of research pertains to attribution methods, which unveil the impact of different input variables on predictions, particularly vital in domains like healthcare and law. These methods aim to provide robust explanations that resonate with human decision-making processes while justifying model responses. Alongside, tests are under development to evaluate the authenticity of explanation methods, ensuring they accurately reflect system workings and resist manipulation. When coupled with formal verification of attributes like fairness and safety, these explanation techniques can provide the ethical and fitting implementation of impactful AI applications. Ultimately, pursuing enhanced interpretability is a central point in preventing trust and responsibly overseeing autonomous systems. Continued research on the evolution and evaluation of explanatory tools can effectively ensure models are employed accountably, reaping their full benefits as discussed in Table 11.1.

Table 11.1: Key areas for improving model performance and responsible usage

Heading	Description
Adapting AI	Examining ways to make AI systems more adaptable, such as experimenting with different data distributions, continuous learning, and self-supervised pre-training.
Interpreting models	Providing tools like attribution techniques, interactive debugging, and counterfactual explanations, focusing on enhancing model interpretability and descriptions.
Optimizing generalization	Improving transfer learning performance, addressing issues outside training contexts, advancing techniques to increase AI model generalization, and more.
Making decisions sense	Investigating how AI models make decisions to gain more information through interactive tools, attribution techniques, and improved decision-making clarity.
Responsible AI	Ensuring ethical AI comprehension and utilization to deliver trustworthy and moral decision-making, particularly in crucial fields like law and healthcare.

Future of AI in Material Science and Construction: Advanced Materials

The field of advanced construction materials is poised to undergo significant change at the nexus of material science and AI. The ability to significantly speed up the discovery and creation process is already being demonstrated by integrating ML and data-driven methodologies. AI's expertise makes fast, high-throughput simulations and characterizations of a wide range of candidate formulations possible, revealing unique combinations for customized attributes far before empirical investigation (Vasudevan et al., 2019a). This ability focuses on materials with particular electrical, thermal, structural, or other customized properties necessary for complex architectural solutions. Additionally, computational methods are crucial in revealing the complex interdependencies between structure, processing, and performance that frequently defy human comprehension (Tătaru et al., 2021). The complexity of material formulation spaces is expertly navigated by AI, telling unexpected synergies and enhancing insights. This, therefore, feeds the development of nanoscale engineering, whether it be the precise carving of ceramic grain boundaries or the creation of polymer chain morphologies.

Computers are ready to assist chemists and materials engineers in quick prototype production as AI-generated insights easily weave into automated synthesis pathways. The linked experiments and live testing are led by AI-trained robotic systems, transforming manual procedures into a continuous, closed-loop refining process. Looking ahead, robotics and AI fusion can create new material compositions based on fundamental concepts. With ML-driven adaptive manufacturing techniques like 4D printing, this vision opens the door to the possibility of materials that are always adaptable to changing architectural requirements. With the addition of cutting-edge materials that redefine performance, sustainability, and resilience benchmarks, this upcoming disruptive innovation has the potential to completely transform infrastructure and position it to take on the constantly evolving issues of the future.

AI-driven Materials Design

AI is revolutionizing how new materials are conceived and developed through computational design methodologies. AI-driven simulation techniques can comb through billions of molecular, atomic, and composite configurations to find combinations that outperform current materials. Before new compounds are physically synthesized, deep learning algorithms have shown they can adequately anticipate features, including crystalline structure, bandgap, density, and stress-strain responses. This makes it possible to find promising candidates for experiments. Enzymatic and biochemical engineering are two fields that will benefit from AI's ability to create optimized protein sequences and small molecule formulations. ML also influences the development of new 2D materials, with algorithms predicting the optical, thermal, and electrical properties of novel 2D metals, polymers, and van der Waals hetero-structures (Yin et al., 2021). AI is also boosting additive manufacturing processes by computationally developing resins, thermoplastics, and photopolymers with appropriate structural characteristics, sustainability profiles, and curing behaviors for 3D printing applications (Himeur et al., 2023). The best AI-proposed formulations can subsequently be manufactured by experimentalists using automated high-throughput techniques like parallel deposition equipment and continuous flow reactors. As a result, AI serves as a potent computational microscope to help us navigate the complex problem of purposely creating matter from the bottom up.

Materials on Demand

The combination of additive manufacturing techniques and AI is writing a new chapter in the history of material creation, according to the innovation pages. The orchestrations of sophisticated robotic production facilities, directed by AI, unfold the capacity to create a wide range of customized substrates on demand, revolutionizing the fundamental nature of fabrication. AI is vital in the construction industry when optimizing site-mixed concrete mixtures. AI plays the role of an architectural maestro by utilizing deep learning algorithms refined on vast datasets covering structural subtleties, environmental conditions, and inventory specifics. It offers advice for precisely calculated mineral and aggregate blends calibrated to achieve the appropriate compressive strengths or self-healing properties. As AI advances into the industrial sphere, it begins to impact 3D printers for metal and directs the development of heat-treated alloys with unique microstructures. Imagine turbines that are resistant to corrosion at extreme temperatures. The biological reactors are controlled by AI, which has given genetic programs to regulate how cellular systems self-assemble into high-performance biocomposites. These AI-controlled living material foundries are master weavers of structural proteins, adhesives, and tissue scaffolds woven to specifications (Sha et al., 2020). In a similar story, AI-driven nanofactories etch ceramic-polymer nanolaminates or structurally superior metal matrixes at the atom level, ushering in a new era for mechanical integrity. The potential for on-demand manufacturing of custom substances is unrestricted by the harmonic interaction of material formulation, architectural design, and subtle processing conditions in this innovative symphony. This catalytic improvement in manufacturing agility transforms critical part production and rapid prototyping, which affects everything from remote locations to space stations. To meet just-in-time needs and shorten design-build cycles, industries are thus cast in a new mold. The story comes to a close with the rise of sophisticated AI, which assumes the role of an administrator for intelligent industrial hubs. A monument to the revolutionary combination of AI and material inventiveness, these hubs act as cradles for synthesizing materials adapted to specific structural and functional requirements.

Construction Applications

The introduction of AI is writing a transforming story in the limitless world of construction, ushering in the era of sophisticated materials that redefine the fundamental nature of building. The development of AI-optimized self-healing materials, such as bacteria-embedded cement that automatically finds and fixes micro-cracks to increase infrastructure lifespan without the burden of expensive maintenance downtime (Saad et al., 2020), is at the core of this revolution. Smart buildings, which seamlessly integrate AI and IoT, emerge as harbingers of sustainability in a symphony of innovation. These buildings raise the bar by continuously balancing internal and external circumstances to optimize energy use, indoor air quality, thermal control, and waste recycling. Here, deep learning algorithms orchestrate a ballet that reduces environmental footprints by delving into vast amounts of building performance data and painstakingly fine-tuning the functionality of intelligent glass, HVAC (heating, ventilation, air conditioning) systems, and more. Even the development of transportable shelters falls inside the purview of AI-guided projects, which is a powerful affirmation of technology's capacity for good. Algorithms skillfully balance design aspects, including ease and speed of assembly/disassembly, material usage, durability, and logistical considerations (Saad, Alaloul, Ammad et al., 2022) across diverse terrains, embodying the resilience and adaptability that are so important in times of disaster. This is accomplished by drawing on lightweight, expandable metamaterials and flat-pack housing prowess (Naser, 2019). AI extends its influence on stimuli-responsive materials in the transportation sector. Shape memory metals, polymers, and composites play the function of adaptable enablers through AI-driven modeling and testing, enabling infrastructure to rearrange dynamically in response to usage demands. Consider how roads or bridges might dynamically change their shapes to coordinate ideal traffic flows during peak hours. However, the story of AI continues. It explores structural bio-composites, demonstrating how AI and the natural environment interact. These living materials weave narratives of sustainability by utilizing cellular self-assembly of protein strands or bacterial cellulose, replicating nature's principles of resilience and regeneration to promote habitats that thrive even under the strain of stress. Construction paradigms are reimagined by the vast tapestry that AI-designed innovative materials weave, giving them a dynamic edge, resilience, and sustainability (Vasudevan et al., 2019b). This vision is realized via continual optimization and self-adaptation, which shapes architectures over decades of use that adapt without effort to changing requirements and fluid environments, forever altering the skyline of possibilities.

Future Prospects

In the coming decades, fully automated AI platforms are expected to be used to create sophisticated metamaterials with superior, multifunctional features. This is because material informatics datasets and AI modeling skills are scaling up quickly (Goyal et al., 2023). AI-driven theory and testing would lead to revolutionary advancements in fields like nanotechnology, robotics, and synthetic biology. The most significant aspect of these breakthroughs is that they may make it possible to produce customized, sustainable materials on demand for a variety of infrastructure projects as well as for industries ranging from transportation, defense, and space construction. Thus, the combination of AI and materials science has the potential to alter engineering technologies and humanity's-built environment fundamentally. AI has created a paradigm change in the dynamic field of materials research and building. The examination of key issues that each significantly influence how materials are created, put to use, and incorporated into construction practices serves to highlight this shift. Table 11.2 discusses a comprehensive overview of key topics of advanced material science and construction in artificial intelligence.

Future of AI in Material Science and Construction: Advanced Manufacturing

As AI assumes center stage and is ready to revolutionize manufacturing processes via the lens of advanced intelligent and data-driven fabrication, a revolution in the construction industry is on the horizon. The world of industrial production is changing, propelled by advances in AI and robots and magnified by the explosion in computing power and data accessibility. Imagine a world of "digital factories" where AI seamlessly integrates into the workflow. Deep learning algorithms examine design schematics and provide insights for ergonomic improvements that reduce the risk of damage (Cankaya et al., 2023). Assembly sequences are verified for effectiveness using computational fluid dynamics simulations controlled by neural networks. Due to intelligent robotics and AI orchestration, precision cutting, welding, finishing, and assembly activities will become a dance directed by intelligent machines (Arinez et al., 2020). The seamless collaboration between humans and robots that results from the combination of language processing and computer vision will usher in an era when borders are blurred. The backbone will be a distributed cloud architecture coordinating additive production cells that create customized components as needed. We now enter the era of cutting-edge 4D printing technologies that make reinforced concrete beams that can change shape in response to structural demands. Nano factories will weave ultra-strong composite laminates layer by layer in parallel with living material bioreactors, producing customized biomaterials. Real-time analytics, digital twins, and advanced sensor networks work harmoniously to continuously improve industrial processes as they lead to this massive transition. With the help of AI, parameters such as tool routes, process conditions, and material utilization will be appropriately recalibrated, boosting productivity, sustainability, and cost-effectiveness (Goyal et al., 2023). The picture becomes clear as predictive maintenance algorithms are prepared to predict breakdowns. AI is ready to usher in a time of networked, self-sufficient, intelligent factories, a sign of mass customization in the construction industry. It becomes quickly possible to deploy infrastructure that is suited to local requirements. The elements of advanced manufacturing converge, transforming construction holistically from atoms to entire cities, thanks to AI managing the seamless union of design, simulation, automation, and analytics.

Table 11.2: Developing a Roadmap for Advanced Materials in Construction and Other Fields

Definition	Importance	Differences	Challenges
Using AI to investigate different chemical and structural potentials for material development computationally, speeding up creation through a combination of predictive modeling and experimental validation.	The computational investigation of AI speeds up the development of new materials more quickly than conventional trial-and-error techniques.	AI's role is to forecast the characteristics and layouts of novel materials, while experimental verification ensures their feasibility.	Access to extensive material databases is essential for making accurate predictions. Complexity must be balanced, and sophisticated AI models are needed for these interactions between material attributes.
Based on specific structural requirements and cost-effectiveness, produce specialized materials on-site using AI-driven additive manufacturing.	Customization: This makes it possible to create materials specifically for specialized purposes.	AI suggests certain concrete mixtures and alloy formulas in adaptive manufacturing.	Algorithm Accuracy: Algorithms must guarantee accurate material characteristics. Resource Optimization: It might take much work to balance material availability and cost-effectiveness.
AI-designed materials can be included in architectural projects to improve resilience, efficiency, and sustainability, creating self-healing structures, energy-efficient structures, and adaptive shelters.	Resilient infrastructure is made possible by the adaptability and disaster resistance of AI materials.	Functional Enhancement: AI materials boost energy efficiency and structural performance.	Integration Difficulty: Integrating AI materials into construction processes seamlessly takes careful preparation. Long-term Performance: Ensuring the longevity and potency of materials infused with AI.
Imagine the creation of multifunctional metamaterials with superior properties across multiple fields due to AI-driven material development.	AI developments will revolutionize industries like nanotechnology and robotics.	Multifunctional metamaterials are created using AI and have a wide range of capabilities.	Scaling Data and Models: Managing the rapid expansion of complicated AI models and material informatics data. Addressing potential hazards and ethical issues related to content produced by AI.

Intelligent Robotics

AI-enabled intelligent robotics' transformational potential for changing the construction industry. These innovative solutions are about to automate everyday manual chores in the sector. Precision welding, delicate coatings, and complex part assembly will be performed by robotic systems with computer vision, decision-making algorithms, and sophisticated manipulation skills under the guidance of AI. Expertly designed with AI, these tireless automated tools perform micron-accurate maneuvers that are difficult for humans to match. Furthermore, by dynamically modeling throughput and spatial layouts and enabling autonomous reorganization in response to changing demand, AI supports the optimization of fabrication lines for maximum efficiency. An innovative method of job site monitoring and maintenance is provided by drones and crawling robots that are powered by computer vision and ML. These portable wonders analyze dangerous constructions in real-time and provide information about defects in the cloud (Bhowmik et al., 2022). The predictive power of AI allows for proactive maintenance scheduling, improving safety, reducing downtime, and streamlining asset management (Bhanji et al., 2021). Intelligent robots are where the transformation will happen, with autonomous, optimized assembly, inspection, and maintenance workflows set to revolutionize the construction industry.

Advanced Additive Manufacturing

The impressive transformation of modern additive manufacturing techniques, such as 3D printing, with the introduction of AI capabilities. Empowered neural networks explore complex 3D scans and CAD designs while utilizing enormous datasets that cover structural models, material parameters, and previous print results (Qureshi et al., 2022). With this expertise, AI provides customized material mixes, whether made of composites, plastic, or concrete, suited for the desired strength and appearance. Aside from that, AI organizes effective print routes, reducing material waste and facilitating the on-site manufacture of fixtures, framework parts, and molds. By enabling AI to create programmable intelligent materials that change shape under specific circumstances, the idea of "4D printing" amplifies this advancement (Tătaru et al., 2021). AI creates novel applications like energy-harvesting façades and flexible formwork by creating microscopic compositions and interior structures. In this phase, customized on-site production, the emergence of configurable intelligent materials, and advanced additive processes are revolutionizing the building industry.

Future of AI in Material Science and Construction: Advanced Construction

The increasing adoption of AI in the construction sector has the potential to change accepted methods. Intelligent technologies are developing due to the integration of robotics, computer vision, ML, and simulation, and they are on the verge of revolutionizing the design, building, and management processes (Saad, Alaloul, Rasheed et al., 2022). AI-driven systems that utilize digital twin models and large databases to plan elaborate projects with previously unattainable precision. Architects can use neural networks to make real-time design revisions that maximize productivity, material use, and compliance. Construction and operation are covered by digital simulations, which automatically fix problems (Sagdic et al., 2022). AI-driven collaborative workflows support data-driven coordination. Based on availability, workload, and job locations, AI-guided dynamic scheduling agents optimize tasks and teams, and computer vision combined with building information models enables remote monitoring and early deviation identification. AI-powered construction promises unmatched versatility, productivity, and resilience with adaptive robotics, self-assembly materials, and IoT-driven infrastructure, defining a future of sustainable buildings completed quickly and affordably with these transformative intelligent technologies.

Intelligent Design and Planning

Architecture transforms as AI develops generative tools that work with human designers to assess various choices. These techniques use massive neural networks refined on enormous databases of award-winning buildings to analyze building designs and suggest real-time improvements quickly. AI becomes the architect's ally as it delves into CAD models, structural calculations, material specifications, and more. It offers cost-effective alterations that balance the need for load-bearing requirements, construction viability, and code compliance. Each iteration of the design interaction between architects and AI improves utility and efficiency. Digital twin modeling supported by AI accelerates the architectural innovation stage (Sha et al., 2020). To identify flaws early on, this transformative phase virtually simulates proposed designs across situations that affect complete lifecycles. Simulations factor in forces like wind, settling, and temperature swings with increased realism due to data from IoT collected from actual buildings. AI creates a frenzy of thousands of simulations each minute, suggesting design improvements that strengthen resilience, reduce maintenance requirements, or reduce environmental footprints. It is possible to forecast the structural reaction to hurricanes, earthquakes, or fires in a single step, complying with building rules and guaranteeing that client safety requirements are fulfilled. This combination of AI-powered generative tools, digital twinning, and data-driven virtual assessments revolutionizes the design process. It speeds up the procedure and reveals optimization opportunities that the naked eye would have missed (Goyal et al., 2023). With AI as their ally, architects can focus on enhancing nuanced aesthetics and experiential components, producing higher-quality, sustainable buildings that align with the philosophy of lowering waste, risk, and costs. AI is the driver ready to transform the core principles of intelligent building design practices.

Adaptive Construction Management

The construction industry is undergoing a significant transformation as it persistently adopts the convergence of cutting-edge technologies, including AI, robotics, automation, and data analytics. The industry is going on a dramatic trajectory powered by innovation and departing from conventional labor-intensive ways, altering its core. The project lifecycle is transformed by AI, which brings remarkable efficiency from the beginning of conceptual planning through the precise orchestration of construction and subsequent facilities management. Generative design enhances the planning landscape by utilizing large datasets to quickly browse possibilities, and digital twin simulations reveal virtual validations of projects (Bhandal et al., 2022). Intelligent scheduling that optimizes timeframes and resource allocation based on past trends is the next movement in the symphony (Oleśków-Szłapka et al., 2019).

Drones connected to AI and equipped with computer vision autonomously supervise construction sites, assuring worker safety and tracking progress. This dynamic supports the optimization of operations. A crucial role is played by the field of robotics, which frees up human hands from dangerous jobs like welding and the delicate choreography of 3D-printed formwork. AI-enhanced exoskeletons allow workers to carry larger loads while fending off tiredness. Through defect identification, computer vision drives automated quality control and maintenance, while predictive algorithms are integrated into the fabric of supply chain management to reduce waste and avoid delays (Qureshi et al., 2021). AI empowers autonomous heavy gear, enabling faster completion of tasks like carefully excavating exact foundations and 3D-printing tunnels (Naser, 2019). AI maps the way towards intelligent retrofitting and repairs, responding to changing needs over time by building on digital records of construction and material usage. The industry is poised for unmatched changes as the convergence of AI, robots, automation, drones, and analytics approaches. Using advanced building methods supported by data-driven intelligent systems will allow for fine-tuning quality, resilience, and sustainability criteria while simultaneously reducing costs and project durations. The fabric that forms the spine of the building infrastructure is prepared to meet social needs with a higher degree of effectiveness and refinement.

Future of AI in Material Science and Construction: Advanced Infrastructure

AI is quickly evolving, as shown by the exponential rise of its computer power, the development of its algorithms, and the accessibility of large digital datasets. An altogether new era of "intelligent infrastructure", marked by the convergence of AI, robotics, sensing technologies, and revolutionary materials, is being created due to this disruptive momentum (Tătaru et al., 2021). The fundamental concepts underlying the design, engineering, and management of our built environments are poised to be redefined by this convergence, giving them the resilience to face emerging challenges. This paradigm places AI at the center of optimizing every aspect of the design of vital infrastructure, powered by realistic computer simulations that seamlessly include current environmental data. The assessment of infrastructure performance under various scenarios is made possible by combining digital twin models with predictive analytics (Li et al., 2022), allowing for proactive strengthening and retrofitting procedures. Using computer vision, IoT sensors, and neural networks to conduct in-the-moment structural inspections and anticipate maintenance needs is accelerating automation.

AI also directs the development of novel self-repairing materials, such as self-healing concrete, which can autonomously recognize and fix structural flaws (A. Ammad et al., 2018). Precision and adaptability-driven robotics are emerging as a transformational force in infrastructure development. Infrastructure develops into autonomous systems that continuously recycle and renew local resources under the guidance of the circularity principles (Wang et al., 2022). Our cities' infrastructure is significantly transformed thanks to a harmonious synthesis of AI, automation, and intelligent ecosystems. They become strong, effective, and adaptable organisms that work with nature to achieve a condition of self-monitoring and self-maintenance while requiring the least amount of human intervention. This harmonization results in infrastructure that serves as a model for overcoming adversity and a beacon of sustainability for future generations.

Intelligent Design of Critical Infrastructure

AI, particularly in risk and resilience analysis, is driving a revolutionary transition in critical infrastructure. With the advent of AI-powered digital twin modeling, established design workflows are being redefined by producing enormous datasets from virtual duplicates of projects coupled with real-time IoT sensor data from existing structures (Rasheed et al., 2022). In this context, AI takes on the role of a conductor, coordinating tens of thousands of physics-based simulations that replicate component performance under various risky circumstances—such as earthquakes, flooding, strong winds, or fires—on an unprecedented scale. Neural networks examine patterns in the simulation results, exposing hidden structural flaws that could have gone unnoticed. As a result of these discoveries, AI presents itself as a wise advisor and suggests exact design improvements and optimizations. For example, it might promote strengthened joints, the use of more corrosion-resistant alloys, or the addition of extra safety measures. Potential dangers are discovered early in the planning process thanks to this meticulous process of virtual analysis. In the background, ML algorithms explore past archives of maintenance data for critical infrastructure like bridges, power plants, and water treatment facilities (Flah et al., 2021). AI skillfully identifies components vulnerable to unanticipated failures using automated pattern recognition among large stores of failure data. The retrofitting recommendations explicitly target systems prone to unreliability, preventing dangerous bottlenecks and service interruptions in the process (Himeur et al., 2023). This tactical move prevents expensive disruptions by ensuring that the infrastructure is resilient despite difficulty. AI is the forerunner of change, armed with data-driven digital tools that seamlessly combine resilience optimization, dependability analysis, and risk modeling. This orchestration ensures that crucial infrastructure survives difficulties and emerges as a pillar, providing communities with unshakeable dependability as intended and enduring the test of time with a steadfast dedication to excellence.

Self-healing Construction

The creation of self-healing infrastructure materials is an example of how AI-driven material innovation has ushered in 'adaptive' building. AI-guided manufactured self-healing cement packed with microscopic capsules holding bacterial spores and chemicals that promote speedy healing is promising for crucial infrastructure like dams and bridges (Tee & Ouyang, 2018). Real-time AI sensor data processing identifies hairline fractures caused by seismic activity and other variables, causing capsules to burst and release their contents. Bacterial spores react with the fillers within hours, precipitating and depositing minerals that fill gaps. Similar to self-reinforcing concrete, AI-optimized 3D printed self-reinforcing concrete reinforces internal linkages damaged by earthquakes automatically using nanoparticles and shape memory polymers. AI-guided intelligent materials embedded in the fractured aggregates contract and reposition them to ensure the correct composition for a seamless repair. The infrastructure receives constant regeneration and reinforcement from these AI-designed adaptive materials, lowering maintenance costs and downtime. The result is a building that can withstand disasters and self-heals and restores, representing a significant advancement in resilient engineering.

References

Alimi, O.A. and Meijboom, R. (2021). Current and future trends of additive manufacturing for chemistry applications: A review. *Journal of Materials Science*, 56(30), 16824–16850.

Ammad, A., Kashif Ur Rehman, S., Saad, S. and Abbas, S. (2018). Assessment of Low Strength Concrete with Destructive & Non-Destructive Testing Methods. *3rd International Conference on Emerging Trends in Engineering, Management and Scineces.* https://www.researchgate.net/profile/Syed-Ammad-2/publication/336140903_Assessment_of_Low_Strength_Concrete_with_Destructive_Non-Destructive_Testing_Method's/links/5d920adca6fdcc2554a70ffe/Assessment-of-Low-Strength-Concrete-with-Destructive-Non-Destructi

Ammad, S., Saad, S., Bashir, M.T., Qureshi, A.H., Altaf, M. and Rasheed, K. (2021). Construction Accidents via Integrating Building Information Modeling

(BIM) with Emerging Digital Technologies: A Review. *2021 3rd International Sustainability and Resilience Conference: Climate Change*. https://doi.org/10.1109/IEEECONF53624.2021.9668066.

Arinez, J.F., Chang, Q., Gao, R.X., Xu, C. and Zhang, J. (2020). Artificial intelligence in advanced manufacturing: Current status and future outlook. *Journal of Manufacturing Science and Engineering*, 142(11), 110804.

Ashebir, D., Pasquini, M. and Bihon, W. (2007). Urban agriculture in Mekelle, Tigray state, Ethiopia: Principal characteristics, opportunities and constraints for further research and development. *Cities*, 24(3), 218–228.

Bag, S., Pretorius, J.H.C., Gupta, S. and Dwivedi, Y.K. (2021). Role of institutional pressures and resources in the adoption of big data analytics powered artificial intelligence, sustainable manufacturing practices and circular economy capabilities. *Technological Forecasting and Social Change*, 163, 120420. https://doi.org/https://doi.org/10.1016/j.techfore.2020.120420.

Bhandal, R., Meriton, R., Kavanagh, R.E. and Brown, A. (2022). The application of digital twin technology in operations and supply chain management: A bibliometric review. *Supply Chain Management: An International Journal*, 27(2), 182–206. https://doi.org/10.1108/SCM-01-2021-0053.

Bhanji, S., Shotz, H., Tadanki, S., Miloudi, Y. and Warren, P. (2021). *Advanced Enterprise Asset Management Systems: Improve Predictive Maintenance and Asset Performance by Leveraging Industry 4.0 and the Internet of Things (IoT)*. https://doi.org/10.1115/JRC2021-58346.

Bhowmik, A., Berecibar, M., Casas-Cabanas, M., Csanyi, G., Dominko, R., Hermansson, K., Palacin, M.R., Stein, H.S. and Vegge, T. (2022). Implications of the BATTERY 2030+ AI-assisted toolkit on future low-TRL battery discoveries and chemistries. *Advanced Energy Materials*, 12(17), 2102698.

Cankaya, B., Topuz, K., Delen, D. and Glassman, A. (2023). Evidence-based managerial decision-making with machine learning: The case of Bayesian inference in aviation incidents. *Omega*, 120, 102906. https://doi.org/https://doi.org/10.1016/j.omega.2023.102906.

Choi, H.-W., Kim, H.-J., Kim, S.-K. and Na, W.S. (2023). An overview of drone applications in the construction industry. *Drones*, 7(8). https://doi.org/10.3390/drones7080515.

Fava, F., Totaro, G., Diels, L., Reis, M., Duarte, J., Carioca, O.B., Poggi-Varaldo, H.M. and Ferreira, B.S. (2015). Biowaste biorefinery in Europe: Opportunities and research & development needs. *New Biotechnology*, 32(1), 100–108.

Flah, M., Nunez, I., Ben Chaabene, W. and Nehdi, M.L. (2021). Machine learning algorithms in civil structural health monitoring: A systematic review. *Archives of Computational Methods in Engineering*, 28(4), 2621–2643. https://doi.org/10.1007/s11831-020-09471-9.

Goyal, P., Kumar, S. and Sharda, R. (2023). A review of the Artificial Intelligence (AI) based techniques for estimating reference evapotranspiration: Current trends and future perspectives. *Computers and Electronics in Agriculture*, 209, 107836.

Himanen, L., Geurts, A., Foster, A.S. and Rinke, P. (2019). Data-driven materials science: Status, challenges, and perspectives. *Advanced Science*, 6(21), 1900808. https://doi.org/https://doi.org/10.1002/advs.201900808.

Himeur, Y., Elnour, M., Fadli, F., Meskin, N., Petri, I., Rezgui, Y., Bensaali, F. and Amira, A. (2023). AI-big data analytics for building automation and management systems: A survey, actual challenges and future perspectives. *Artificial Intelligence Review*, 56(6), 4929–5021.

Hippalgaonkar, K., Li, Q., Wang, X., Fisher, J.W., Kirkpatrick, J. and Buonassisi, T. (2023). Knowledge-integrated machine learning for materials: Lessons from gameplaying and robotics. *Nature Reviews Materials*, 8(4), 241–260. https://doi.org/10.1038/s41578-022-00513-1.

Li, X., Liu, H., Wang, W., Zheng, Y., Lv, H. and Lv, Z. (2022). Big data analysis of the Internet of Things in the digital twins of smart city based on deep learning. *Future Generation Computer Systems*, 128, 167–177. https://doi.org/https://doi.org/10.1016/j.future.2021.10.006.

López, C. (2023). Artificial Intelligence and advanced materials. *Advanced Materials*, 35(23), 2208683.

Lundvall, B.-Å., Vang, J., Joseph, K.J. and Chaminade, C. (2009). *Bridging Innovation System Research and Development Studies: Challenges and Research Opportunities*. Lund University. https://vpn.aau.dk>files>Bridging-innovation...PDF.

Naser, M.Z. (2019). Properties and material models for modern construction materials at elevated temperatures. *Computational Materials Science*, 160, 16–29.

Niu, P.-H., Zhao, L.-L., Wu, H.-L., Zhao, D.-B. and Chen, Y.-T. (2020). Artificial intelligence in gastric cancer: Application and future perspectives. *World Journal of Gastroenterology*, 26(36), 5408.

Oleśków-Szłapka, J., Wojciechowski, H., Domański, R. and Pawłowski, G. (2019). Logistics 4.0 maturity levels assessed based on GDM (grey decision model) and artificial intelligence in logistics 4.0-trends and future perspective. *Procedia Manufacturing*, 39, 1734–1742.

Pan, Z., Yu, W., Yi, X., Khan, A., Yuan, F. and Zheng, Y. (2019). Recent progress on generative adversarial networks (GANs): A survey. *IEEE Access*, 7, 36322–36333. https://doi.org/10.1109/ACCESS.2019.2905015.

Qureshi, A.H., Alaloul, W.S., Manzoor, B., Saad, S., Alawag, A.M. and Alzubi, K.M. (2021). Implementation Challenges of Automated Construction Progress Monitoring Under Industry 4.0 Framework Towards Sustainable Construction. *2021 Third International Sustainability and Resilience Conference: Climate Change*, 322–326. https://doi.org/10.1109/IEEECONF53624.2021.9668074.

Qureshi, A.H., Alaloul, W.S., Murtiyoso, A., Hussain, S.J., Saad, S. and Oad, V.K. (2022). Evaluation of 3D Model of Rebar for Quantitative Parameters. *The International Archives of the Photogrammetry, Remote Sensing and Spatial Information Sciences*, XLVIII-2/W, 215–220. https://doi.org/10.5194/isprs-archives-XLVIII-2-W1-2022-215-2022.

Rasheed, K., Saad, S., Shahzad, L., Ammad, S., Ali, A. and Badshah, I. (2021). Application of Gesture Data Recognition in a Human-Interactive Leap Motion Sensor Chair. *2021 International Conference on Data Analytics for Business and Industry, ICDABI 2021*. https://doi.org/10.1109/ICDABI53623.2021.9655819.

Rasheed, K., Shahzad, L., Saad, S., Khan, H.A., Ahmed, W. and Sadiq, T. (2021). Parking Guidance System Using Wireless Sensor Networks. *2021 International Conference on Decision Aid Sciences and Application, DASA 2021*. https://doi.org/10.1109/DASA53625.2021.9682213.

Rasheed, K., Ammad, S., Said, A.Y., Balbehaith, M., Oad, V.K. and Khan, A. (2022). Application of Smart IoT Technology in Project Management Scenarios. *2022 International Conference on Data Analytics for Business and Industry, ICDABI 2022*. https://doi.org/10.1109/ICDABI56818.2022.10041674.

Saad, S., Salah Alaloul, W., Ammad, S., Hannan Qureshi, A., Mohsen Mohammed Alawag, A., Kumar Oad, V. and Altaf, M. (2020). A noded carbon emission tool (NCET) to measure embodied carbon (EC) in compositional cementitious materials. *Solid State Technology*, 63(6), 4099–4105.

Saad, S., Alaloul, W.S., Ammad, S. and Qureshi, A.H. (2022). A qualitative conceptual framework to tackle skill shortages in offsite construction industry: A scientometric approach. *Engineering, Construction and Architectural Management*, 29(10), 3917–3947. https://doi.org/10.1108/ECAM-04-2021-0287.

Saad, S., Alaloul, W.S., Rasheed, K. and Ammad, S. (2022). Modeling and simulation of construction cyber-physical systems. *In: Cyber-Physical Systems in the Construction Sector*, pp. 88–110. CRC Press.

Sagdic, K., Eş, I., Sitti, M. and Inci, F. (2022). Smart materials: Rational design in biosystems via artificial intelligence. *Trends in Biotechnology*.

Sha, W., Guo, Y., Yuan, Q., Tang, S., Zhang, X., Lu, S., Guo, X., Cao, Y.-C. and Cheng, S. (2020). Artificial Intelligence to power the future of materials science and engineering. *Advanced Intelligent Systems*, 2(4), 1900143. https://doi.org/10.1002/AISY.201900143.

Tătaru, O.S., Vartolomei, M.D., Rassweiler, J.J., Virgil, O., Lucarelli, G., Porpiglia, F., Amparore, D., Manfredi, M., Carrieri, G. and Falagario, U. (2021). Artificial intelligence and machine learning in prostate cancer patient management—Current trends and future perspectives. *Diagnostics*, 11(2), 354.

Tee, B.C.K. and Ouyang, J. (2018). Soft electronically functional polymeric composite materials for a flexible and stretchable digital future. *Advanced Materials*, 30(47), 1802560.

Vasudevan, R.K., Choudhary, K., Mehta, A., Smith, R., Kusne, G., Tavazza, F., Vlcek, L., Ziatdinov, M., Kalinin, S.V. and Hattrick-Simpers, J. (2019a). Materials science in the artificial intelligence age: High-throughput library generation, machine learning, and a pathway from correlations to the underpinning physics. *MRS Communications*, 9(3), 821–838. https://doi.org/10.1557/mrc.2019.95.

Wang, Z., Yang, W., Liu, Q., Zhao, Y., Liu, P., Wu, D., Banu, M. and Chen, L. (2022). Data-driven modeling of process, structure and property in additive manufacturing: A review and future directions. *Journal of Manufacturing Processes*, 77, 13–31.

Yin, H., Sun, Z., Wang, Z., Tang, D., Pang, C.H., Yu, X., Barnard, A.S., Zhao, H. and Yin, Z. (2021). The data-intensive scientific revolution occurring where two-dimensional materials meet machine learning. *Cell Reports Physical Science*, 2(7), 100482. https://doi.org/10.1016/j.xcrp.2021.100482.

CHAPTER

12

The Conclusive Role of AI in Industry 4.0 Construction Materials

Syed Saad, Kumeel Rasheed, Syed Ammad, Zawar Ali and Ahmad Zaland

Summary of the Key Takeaways

This book has given a thorough review of the current artificial intelligence (AI) revolution and its wide-ranging effects on the worldwide construction industry. Exploratory research into machine learning (ML) and neural networks that started decades ago has evolved into a full-fledged digital transformation that is still going strong (Lindsay, 2019). It is abundantly clear from the chapters exploring topics ranging from applications of materials science to the modernization of urban infrastructure using big data techniques that AI is poised to fundamentally alter how we plan, design, construct, and maintain the built environment, from the smallest residential projects to the most elaborate engineering feats. As the introduction explained, the construction industry has historically relied on incremental advancements brought about by mechanization and standardized procedures, but it is currently at a turning point where fundamental business models and workflows are being redefined as a result of new digital technologies (Saltz et al., 2018). In terms of its all-encompassing impact, the current paradigm change is analogous to the electrification of industry in the late-nineteenth century. Just as before, if not properly understood and guided towards inclusive outcomes that uplift mankind, such revolutionary change at the societal level is accompanied by both promise and hazard. Our analysis of AI's significant contributions to computer-aided material discovery and design shows that completely new classes of stronger, lighter, and more sustainable building materials with never-before-seen features will be produced using computational design methods. The fusion of these technologically advanced building materials with the robotics, sensors, autonomous vehicles, and digitally integrated project delivery techniques covered in the following chapters paves the way for a new era of precise, ultra-efficient building that can address major issues like energy consumption and housing shortages (Design, 2019). At the same time, as automation spreads across prefabrication yards and construction sites thanks to collaborative robots, exoskeletons, blockchain-verified worker skill certifications, and AI computer vision systems, caution must be taken to respect human laborers as valuable partners rather than as assets to be marginalized or discarded. The social compact may be kept to make sure no member of society feels left behind or homeless throughout industry upheaval through socially conscious legislation, ongoing skills retraining programs, and negotiated transition agreements between unions and businesses generating new hybrid digital roles.

Additionally, as Industry 4.0's digital twins, geospatial systems, and sensor networks optimize and revitalize ageing urban infrastructure with self-healing materials and AI-managed mobility networks, regulatory oversight and community involvement are required to go along with these game-changing tech initiatives (Li, 2022). Residents should be involved in identifying objectives for fair access, environmental protection, and cultural preservation even as economic indicators rise. They also deserve transparency into algorithmic decision-making that guides planning, resource allocation, and disaster response. By managing public areas and infrastructure with data-informed stewardship, "Smart Cities" can fulfill its promise to improve quality of life indicators for all residents rather than just users or environmental variables (Andras et al., 2020). The construction industry is poised to either use AI to pioneer a more equitable, sustainable, and prosperous built environment or risk deepening current societal divisions and environmental issues if innovation loses sight of its greater purpose, as the final chapter highlighted. This industry is well-positioned to demonstrate to the rest of the world how emerging technologies can and should be used for the benefit of people and the well-being of the planet as a whole (Wong et al., 2021). This is done by placing a higher priority on ethics, inclusion, stakeholder collaboration, and globally coordinated progress assessment rather than competitive pressures or specific profit-driven goals. Future built environments will be characterized by shared prosperity, accessibility, and harmony between people and places if the opportunities and challenges presented throughout this book are addressed with vision, wisdom, and care for future generations. This will be made possible by wise, responsibly applied digital transformation.

Future Perspectives and Trends in AI in Material Science and Construction

In ways that were previously only possible in science fiction, AI is set to revolutionize the domains of material science and building. Both material scientists and builders are on the verge of building creative solutions at an unprecedented scale thanks to breakthroughs in AI-driven technologies, including accelerated ML technology and pervasive sensor networks generating massive datasets (Jimenez-Gomez et al., 2020). Computational materials design is where AI will have the biggest influence since it uses supercomputers and ML to help find unique material structures that are optimized right down to the atomic level. Beyond focusing on particular material characteristics, generative AI models may even develop completely new theoretical frameworks for materials, leading to self-assembling nano-engineered composites with hierarchical architectures, programmable materials, and self-healing mechanisms at various scales (Nations, 2015). AI will also enable the creation of specialized metamaterials, including invisibility cloaks and acoustic metamaterials, with unique capabilities that go well beyond those of conventional materials.

Through digital twin modelling, AI will simultaneously bring in an era of intelligent, anticipatory infrastructure. Virtual worlds will be built using real-time construction data from tools like autonomous robotics and nanoscale sensors to accurately replicate physical structures. The pre-construction planning, iterating designs, and optimizing simulations before the actual work begins will be revolutionized by these digital twins (Saad et al., 2021). AI supervision of autonomous machines and real-time feedback for optimized sequencing will improve on-site building (Ramos et al., 2018). Once built, digital twins will act as operating manuals that can be updated as needed, utilizing sensor networks to identify changes and informing maintenance schedules and repair procedures. Entire metropolitan systems will be integrated into data-rich models, enabling planners to predict effects and proactively replace deteriorating infrastructure (Gassmann et al., 2019). Buildings will be dynamic, intelligent systems that change as a result of collective machine intelligence, ultimately bringing benefits to humans in the form of solutions that are accessible, resilient, and sustainable. AI gives the chance to sculpt a constructed environment that advances and adapts naturally for decades to come with responsible development and ethical considerations.

AI Unlocks an Era of Hyper-advanced Materials

Through the use of AI and computational materials design, our world is poised to undergo a significant transformation. These advancements are no longer exclusive to science fiction, but rather, they are becoming a reality. We are navigating complexity at the atomic scale using simulations with billions of parameters, exascale supercomputing capacity, and ML approaches, allowing us to customize material properties down to the molecular level. By moving beyond optimization to create brand-new material design frameworks, this data-driven technique uncovers innovative material structures. Self-assembling nanomaterials with hierarchical designs have the potential to revolutionize infrastructure and transportation by outperforming steel in strength, while being lighter than aluminium (Guo et al., 2018). Semiconducting nanoparticles, DNA origami, or composite microstructures may all lead to the emergence of programmable matter, which dynamically modifies attributes through external stimuli.

Thanks to well-planned embedded mechanisms inside the materials themselves, the future presents the potential of autonomous repair on a scale hitherto unattainable. These systems, which include internal actuators and damage detecting sensors, have the ability to make molecular repairs like re-bonding damaged atomic connections. Large fissures may eventually self-heal on a broader scale without the need for outside assistance. Self-assembling nanostructures, such as those containing healing chemicals that are delivered by implanted nanorods, could be used to accomplish this, enabling materials to independently recover their integrity (Barbosa et al., 2017). With the help of this development, constructions in risky locations, like offshore energy platforms, could be much more resilient, allowing materials to persist for centuries without requiring human upkeep. The spread of metamaterials is another transformational force that is imminent. These composite materials, which were created using cutting-edge manufacturing methods, like 3D printing, display characteristics that go beyond those of their individual components. The possibilities are endless, from impact-absorbing hyperelastic textiles to invisibility cloaks that manipulate light to topological insulators with unique electron behaviors that inspire quantum technologies. While offering customized mechanical qualities that are optimized for particular applications, intelligent composite design, guided by AI, promises to lower energy intensity. By identifying damage signals and guiding repair tactics, integrated sensing capabilities will enable predictive maintenance of ageing infrastructure (Jarrahi et al., 2023). Additionally, AI-assisted computational biomimicry has the potential to reveal the final mysteries of nature's flawlessly optimized systems. Dynamic cellular automata modeling initiatives that imitate the microstructures of bone, shell, and tissue may result in ultra-lightweight, impact-resistant material architectures that are appropriate for automobiles, sturdy housing, or cutting-edge prosthetics. Computational creativity has the ability to create totally new hybrid organic-synthetic systems in addition to simulating the biological systems. Through resource informatics, optimized renewable energy technologies, and computational catalysis, AI is also positioned to transform sectors of the economy that rely on raw materials. AI-accelerated engineering could hasten the creation of effective synthetic pathways for green chemistry, fusion power, or renewable energy sources at a scale that has never been achieved (Forum, 2016). This would help to supplement finite resources while reducing emissions. Circular manufacturing paradigms will be made possible by the adoption of closed-loop extraction methods, made possible by the tracking of valuable material flows inside global digital product passports. Machine intelligence will fundamentally alter the design landscape at the molecular level, influencing the sophisticated materials of a sustainable future.

Digital Twins Usher in an Era of Living, Intelligent Infrastructure

The widespread use of digital twin modeling, which includes all stages from pre-design to long-term operations and maintenance, is causing a dramatic transformation in the building sector. Massive amounts of real-time, multidimensional data from sources like drones, nanoscale sensors, computer vision analysis of video taken on construction sites, and automated surveying are powering this transformation and allowing for the creation of incredibly detailed virtual environments that replicate physical infrastructure systems down to the molecular level (Forum, 2016). These "living sandboxes" provide as testing grounds for iterative design improvement, taking into account stakeholder preferences, geological conditions, and environmental implications as determined by computational fluid dynamics. Before any physical building takes place, autonomous virtual construction worker simulations can prototype, examine 3D-printed materials, determine the best sequencing and safety procedures, and fine-tune plans. With the help of AI, this immersive virtual pre-construction planning guarantees unmatched precision and dependability in labor, cost, and schedule projections. AI executes digitally planned build sequences on the construction site, coordinating and continuously optimizing fleets of completely autonomous machines and robotic laborers while being guided by real-time sensor readings that improve underlying BIM/CIM models (Saltz et al., 2018). AI simultaneously oversees just-in-time deliveries, personnel allocation based on occupancy requirements, and constant financial performance monitoring, eradicating expensive delays and overruns through self-correcting virtual contingencies tested within simulations. After construction is finished, digital twins progress from being simple instructions to being user interfaces for interactive living systems. Through continuous learning and autonomous optimization not possible within isolated physical structures, sensor networks detecting even microscopic structural changes inform AI-generated maintenance forecasts and self-diagnosing repair protocols, aiming to extend asset lifespans far beyond conventional designs.

The design and administration of infrastructure systems are going to undergo a major shift as a result of the integration of digital modelling at the urban scale. AI planners obtain a unique insight of complex urban dynamics through extensive computer simulations that cover interconnected networks of roadways, utilities, buildings, and public works. Regional digital twins, which replicate geographic contexts right down to the street level, make it possible to conduct evaluations across a variety of scenarios and time frames, which was previously not achievable with isolated project-level research (Lindsay, 2019). It is possible to systematically test a variety of factors for their effects on system performance, resilience, and efficiency (Rasheed et al., 2021). These factors include suggested public policies, expected climate events based on geolocation, and population migration patterns modeled from socioeconomic situations. With the aid of simulation-based foresight, this skill enables the proactive upgrade of deteriorating municipal assets. With more confidence, decisions about renewal techniques, including whether to adapt dated water distribution pipes or to put in place localized modular replacements, may be made. The use of integrated lidar sensor networks on autonomous drones to generate hyper-localized 3D environmental mapping results in very detailed inputs that improve emergency response logistics and community resilience planning. Parallel to this, quantitative sustainability analyses measure the waste reductions that can be achieved via the application of prefabrication techniques and alternative material selections at the size of entire developments across decades (Andras et al., 2020). Digital twins optimize parameters like embodied carbon and water usage by modelling entire lifecycles from raw material extraction to use and disposal, driving procurement regulations that support circular economies. According to the concept of living infrastructure, built environments continuously change and grow as dynamic extensions of collective intelligence, continuously learning from massive datasets. Through computational design tools that integrate knowledge to optimize density, accessibility, resilience, and well-being, cities themselves are envisioned to emerge. According to long-term estimates that balance priorities, integrated planning places land use, transportation systems, green spaces, and other variables in their appropriate locations. Infrastructure is designed to have service lifetimes that are comparable to those found in nature, thanks to distributed intelligence enabled by digital twins. While fluid optimization of effectiveness, safety, and occupant experience realizes infrastructure's full potential for dynamic coevolution with its urban setting, molecular-scale fault diagnosis and regenerative material mechanisms become the norm. Metropolitan-scale digital twins connected globally spread insights worlwide, accelerating continuous learning across all projects. AI trained through examples of both triumphs and failures, maintains physical and virtual networks on its own as they progress in harmony. The reproduction of optimized living designs that make use of material digitalization is replacing the use of discrete projects in building (Ammad et al., 2021). Instead of being static blueprints, infrastructure effectively exists through machines that advance shared structures (Design, 2019). Sentient modeling platforms create communities that are enhanced by collective intelligence, reshaping civilization as a result of synthetic realities that are spatially aware and coevolve with nature.

References and Further Readings

The research offered in this book is based on a wide range of sources that show how quickly developments in AI, digital technology, and the construction sector are developing. The following references include key research programs, reports, case studies, and publications that have helped form our knowledge of AI's expanding role in reinventing how we design, create, and maintain the built environment, however space limitations restrict exhaustive citation (Jimenez-Gomez et al., 2020). The World Economic Forum, McKinsey Global Institute, Autodesk, and World Bank have all produced foundational industry roadmaps and programs that offer thorough perspectives on the factors driving the digitization of the US $10 trillion construction sector. Experts have been gathered by organizations like the Royal Academy of Engineering to outline responsible innovation guidelines for technological applications in infrastructure and building. National AI policies from the US, UK, Canada, and Singapore, among others, outline guiding principles for workforce development, R&D (research and development) spending, and promoting cooperation between governments, businesses, and academia. Key publications from the Construction Industry Institute, Brookings Institute, Boston Consulting Group, and associations that represent heavy equipment manufacturers are among the primary sources for information on AI and robots automating on-site operations (Wong et al., 2021). Case studies by top construction companies and technology vendors highlight actual pilot projects modernizing techniques including pre-casting, welding, and finishing. Databases of robotics and automation vendors document the expanding product ecosystem, which ranges from exoskeletons to collaborative robots. It was necessary to consume materials from Annual Reviews, Elsevier, Springer, RSC Publishing, and ACS on themes ranging from tissue engineering, autonomous repair, and sustainability to understand advanced materials research utilizing computational design and AI (Ramos et al., 2018). Contemporary insights were presented in conference proceedings from organizations including the American Institute of Physics, Materials Research Society, and International Materials Reviews.

Platform documentation and vendor solution databases clarified new tools. Cross-referencing information and standards organizations on smart cities, the Internet of Things, sensor networks, and predictive analytics applied to transit, utilities, public spaces, and more is necessary for monitoring and managing infrastructure systems with AI. Specific problem areas, such as structural health, traffic flows, and resource optimization, were covered in journal papers from Elsevier, IEEE, and MDPI. In terms of the socioeconomic aspects of AI, the World Bank, OECD, ITU, and United Nations all have key frameworks for evaluating fair and sustainable effect that are in line with the SDGs (Nations, 2015). Through organizations that represent construction unions globally, stakeholder perspectives that are labor-focused arise. IEEE, ACM, and other computer societies provide ethical guidelines that establish technological accountability and governance best practices that are closely followed. The aforementioned is simply a high-level sample of the extensive ongoing research being conducted by numerous organizations that is improving our collective understanding of the promise and difficulties of AI in transforming the physical world we live in (Guo et al., 2018). To make sure AI realizes its potential to elevate humanity through intelligent infrastructure, design, and construction, ongoing monitoring and debate of new research advancements, demonstration projects, policy discussions, and emerging technologies through various channels will be essential. Here are some of the references for the further readings.

Additional Resources for Further Learning

1. *Application of Artificial Intelligence in New Materials Discovery* by Inamuddin, Maha Khan, Jafar Mazumder.

2. *Design of Infrastructure Materials: The Genome Approach* by Linbing Wang, Wenjuan Sun, Yue Hou.
3. *Artificial Intelligence in Industry 4.0* by Alexiei Dingli, Christina Klüver, Foaad Haddod.
4. *Machine Learning and Data Science in the Oil and Gas Industry* by Patrick Bangert.
5. *Computational Materials Science: An Introduction* by June Gunn Lee.
6. *Artificial Intelligence for Materials Discovery* by Don Monroe.
7. *Materials Science and Engineering: Concepts, Methodologies, Tools, and Applications* (Handbook) edited by Information Resources Management Association.
8. *Machine Learning in Materials Science* by Keith T. Butler, Felipe Oviedo, Pieremanuele Canepa.
9. *Computational Materials Science: From Ab Initio to Monte Carlo Methods* by Kaoru Ohno.
10. *Materials Design and Applications* edited by Lucas F.M. da Silva, Paula M.S. Figueiredo, Rui Tojo.
11. *Machine Learning in Chemistry: The Impact of Artificial Intelligence* edited by Pablo M. Piaggi, Andreas Singraber.
12. *Artificial Intelligence for Big Data: Complete Guide to Automating Big Data Solutions Using Artificial Intelligence Techniques* by Anand Deshpande.
13. *Artificial Intelligence for Materials Science* by Yuan Cheng, Tian Wang, Gang Zhang.
14. *Machine Learning for Dummies* by John Paul Mueller, Luca Massaron.
15. *Reconstructing our Orders: Artificial Intelligence and Human Society* Publisher: Springer; Shanghai University Press, Singapore, Shanghai, China, 2019.
16. *Industry 4.0, AI, and Data Science* edited by Vikram Bali, Kakoli Banerjee, Narendra Kumar, Sanjay Gour, Sunil Kumar Chawla.
17. *Sustainable Nanomaterials for the Construction Industry* by Ghasan Fahim Huseien, Kwok Wei Shah.
18. *Management for the Construction Industry* by Stephen D. Lavender.
19. *Construction 4.0: An Innovation Platform for the Built Environment* by Anil Sawhney, Michael Riley, Javier Irizarry.
20. *Robotics and Automation in Construction* by Carlos Balaguer, Mohamed Abderrahim.
21. *Artificial Intelligence in Construction Engineering and Management* by Limao Zhang, Yue Pan, Xianguo Wu, Mirosław J. Skibniewski.

References

Forum, W.E. (2016). Shaping the future of construction: A breakthrough in mindset and technology. *World Economic Forum*, 1–61.

Ammad, S., Saad, S., Bashir, M.T., Qureshi, A.H., Altaf, M. and Rasheed, K. (2021). Construction Accidents via Integrating Building Information Modelling (BIM) with Emerging Digital Technologies: A Review. *2021 3rd International Sustainability and Resilience Conference: Climate Change*. https://doi.org/10.1109/IEEECONF53624.2021.9668066.

Andras, I., Mazzone, E., van Leeuwen, F.W.B., De Naeyer, G., van Oosterom, M.N., Beato, S., Buckle, T., O'Sullivan, S., van Leeuwen, P.J. and Beulens, A. (2020). Artificial intelligence and robotics: A combination that is changing the operating room. *World Journal of Urology*, 38, 2359–2366.

Barbosa, F., Woetzel, J. and Mischke, J. (2017). *Reinventing Construction: A Route of Higher Productivity*. McKinsey Global Institute.

Design, E.A. (2019). A vision for prioritizing human well-being with autonomous and intelligent systems. *Report: The IEEE Global Initiative for Ethical Considerations in Artificial Intelligence and Autonomous Systems, 2018.*

Gassmann, O., Böhm, J. and Palmié, M. (2019). *Smart Cities: Introducing Digital Innovation to Cities*. Emerald Publishing Limited.

Guo, K., Lu, Y., Gao, H. and Cao, R. (2018). Artificial intelligence-based semantic internet of things in a user-centric smart city. *Sensors*, 18(5), 1341.

Jarrahi, M.H., Askay, D., Eshraghi, A. and Smith, P. (2023). Artificial intelligence and knowledge management: A partnership between human and AI. *Business Horizons*, 66(1), 87–99.

Jimenez-Gomez, C.E., Cano-Carrillo, J. and Lanas, F.F. (2020). Artificial intelligence in government. *Computer*, 53(10), 23–27.

Li, L. (2022). Reskilling and upskilling the future-ready workforce for industry 4.0 and beyond. *Information Systems Frontiers*, 1–16.

Lindsay, J. (2019). Building capacity for global connections and collaborations – New perspectives. *Synergy*, 17(1).

Nations, U. (2015). *Transforming our World: The 2030 Agenda for Sustainable Development*. New York: United Nations, Department of Economic and Social Affairs.

Ramos, F., Trilles, S., Muñoz, A. and Huerta, J. (2018). Promoting pollution-free routes in smart cities using air quality sensor networks. *Sensors*, 18(8), 2507.

Rasheed, K., Saad, S., Shahzad, L., Ammad, S., Ali, A. and Badshah, I. (2021). Application of Gesture Data Recognition in a Human-Interactive Leap Motion Sensor Chair. *2021 International Conference on Data Analytics for Business and Industry, ICDABI 2021*. https://doi.org/10.1109/ICDABI53623.2021.9655819.

Saad, S., Alaloul, W.S., Shafiq, N., Ammad, S., Qureshi, A.H. and Mohammed, A.A.M. (2021). Visually Programming Automated Slab Positioning Tool (ASPT) using Evolutionary Solvers. *2021 International Conference on Data Analytics for Business and Industry (ICDABI)*, pp. 527–530. https://doi.org/10.1109/ICDABI53623.2021.9655914.

Saltz, J.S., Dewar, N.I. and Heckman, R. (2018). Key concepts for a data science ethics curriculum. *Proceedings of the 49th ACM Technical Symposium on Computer Science Education*, pp. 952–957.

Wong, T.K.M., Man, S.S. and Chan, A.H.S. (2021). Exploring the acceptance of PPE by construction workers: An extension of the technology acceptance model with safety management practices and safety consciousness. *Safety Science*, 139, 105239.

Index

D

E

For Product Safety Concerns and Information please contact our EU representative GPSR@taylorandfrancis.com
Taylor & Francis Verlag GmbH, Kaufingerstraße 24, 80331 München, Germany

www.ingramcontent.com/pod-product-compliance
Lightning Source LLC
LaVergne TN
LVHW010547110826
845149LV00003B/584